Generalized Harmonic Analysis and *Tauberian Theorems*

NORBERT WIENER

Generalized Harmonic Analysis
and
Tauberian Theorems

THE M.I.T. PRESS
Massachusetts Institute of Technology
Cambridge, Massachusetts, and London, England

Generalized Harmonic Analysis and *Tauberian Theorems* first appeared in book form in *Selected Papers of Norbert Wiener*, published jointly in 1965 by Society for Industrial and Applied Mathematics and the M.I.T. Press.
Generalized Harmonic Analysis is here reprinted again by courtesy of the Institut Mittag-Leffler, and *Tauberian Theorems* by courtesy of the *Annals of Mathematics*.

First M.I.T. Press Paperback Edition, August 1966

Library of Congress Catalog Card Number: 66-25164
Printed in the United States of America

Generalized Harmonic Analysis and *Tauberian Theorems*

GENERALIZED HARMONIC ANALYSIS[1]

Reprinted from *Acta Math.*, 55, 1930, pp. 117-258. (Courtesy of the Institut Mittag-Leffler.)

Introduction.

Generalized harmonic analysis represents the culmination and combination of a number of very diverse mathematical movements. The theory of almost periodic functions finds its precursors in the theory of Dirichlet series, and in the quasiperiodic functions of Bohl and Esclangon. These latter, in turn, are an answer to the demands of the theory of orbits in celestial mechanics; the former take their origin in the analytic theory of numbers. Quite independent of the regions of thought just enumerated, we have the order of ideas associated with the names of Lord Rayleigh, of Gouy, and above all, of Sir Arthur Schuster; these writers concerned themselves with the problems of white light, of noise, of coherent and incoherent sources. More particularly, Schuster was able to point out the close analogy between the problems of the harmonic analysis of light and the statistical analysis of hidden periods in such scientific data as are common in meteorology and astronomy, and developed the extremely valuable theory of the periodogram. The work of G. I. Taylor on diffusion represents another valuable anticipation of theories here developed, from the standpoint of an applied mathematician. of the British school, with preoccupations much the same as those of Schuster.

The work of Hahn seems to have a much more definitely pure-mathematics motivation. To the pure mathematician in general, however, and the worker in real function theory in particular, we owe, not so much the setting of our problem, as the chief tool in its attack: the famous theorem of Plancherel, the proof of which Titchmarsh has extended and improved.

It may seem a little strange to the reader that the present paper should contain yet another proof of this much proved theorem. In view, however, of the centralness of the Plancherel theorem in all that is to follow, and more expecially of the fact that the proof here given furnishes an excellent introduction to the meaning and motivation of our proofs in more complicated cases, it has seemed worth while to prove the Plancherel theorem in full.

The germs of the generalized harmonic analysis of this paper are already in the work of Schuster, but only the germs. To make the Schuster theory assume a form suitable for extension and generalization, a radical recasting is necessary. This recasting brings out the fact that the expression

$$\varphi(x) = \lim_{T \to \infty} \frac{1}{2T} \int_{-T}^{T} f(x+t) \bar{f}(t) \, dt \tag{0. 1}$$

plays a fundamental part in Schuster's theory, as does also

$$S(u) = \frac{1}{2\pi} \int_{-\infty}^{\infty} \varphi(x) \frac{e^{iux} - 1}{ix} \, dx. \tag{0. 2}$$

Accordingly, section 3 is devoted to the independent study of these two expressions, and to the definition of $S(u)$ under appropriate assumption as the *spectrum* of $f(x)$.

There are some interesting relations between the total spectral intensity of $f(x)$ as represented by $S(u)$ and the other expressions of the theory. Some of these demand for their proper appreciation a mode of connecting various weighted means of a positive quantity. The appropriate tool for this purpose is the general theory of Tauberian theorems developed by the author and applied to these problems by Mr. S. B. Littauer.

These latter Tauberian theorems enable us to correlate the mean square of the modulus of a function and the »quadratic variation» of a related function which determines its harmonic analysis. The theory of harmonic analysis here indicated has been extended by Bochner to cover the case of very general functions behaving algebraically at infinity. A somewhat similar theory is due to Hahn, who is, however, more interested in questions of ordinary convergence than in those clustering about the Parseval theorem.

The theory of generalized harmonic analysis is itself capable of extension in very varied directions. Mr. A. C. Berry has recently developed a vectorial extension of the theory to n dimensions, while on the other hand, the author himself has extended the theory to cover the simultaneous harmonic analysis of a set of functions and the notions of coherent and incoherent sources of light. A third extension depends on the replacement of the translation group, fundamental in all harmonic analysis, by another group.

To prove that the theory is not vacuous and trivial, it is of importance to give examples of different types of spectra. We do this, both by direct methods, and by methods involving an infinite series of choices between alternatives of equal probability. The latter method, of course, involves the assumption of the

Zermelo axiom: on the other hand, it yields a most interesting probability theory of spectra. This theory may be developed to cover the case where the infinite sequence of choices is replaced by a haphazard motion of the type known as Brownian.

The spectrum theory of the present paper has as one very special application the theory of almost periodic functions. It is not difficult to prove that the spectrum of such a function contains a discrete set of lines and no continuous part, and to deduce from this, Bohr's form of the Parseval theorem. The transition from the Parseval theorem to the Weierstrassian theorem that it is possible to approximate uniformly to any almost periodic function by a sequence of trigonometrical polynomials follows essentially lines laid down by Weyl, though it differs somewhat in detail.

Besides the well-known generalizations of almost periodic functions due to Stepanoff, Besicovitch, Weyl, and the author, there is the little explored field of extensions of almost periodic functions containing a parameter. These have been used by Mr. C. F. Muckenhoupt to prove the closure of the set of the Eigenfunktionen of certain linear vibrating systems. This is one of the few applications of almost periodic functions of a fairly general type to definite mathematicophysical problems. Our last section is devoted to this, and to related matters.

CHAPTER I.

1. Plancherel's theorem.

Plancherel's theorem reads as follows: *Let $f(x)$ be quadratically summable over $(-\infty, \infty)$ in the sense of Lebesgue — that is, let it be measurable, and let*

$$\int_{-\infty}^{\infty} |f(x)|^2\, dx \tag{1.01}$$

exist and be finite. (i.e. $f \subset L_2$). *Then*

$$g(u) = \underset{A\to\infty}{\mathrm{l.i.m.}}\ \frac{1}{\sqrt{2\pi}} \int_{-A}^{A} f(x) e^{iux}\, dx \tag{1.02}$$

(*where* l.i.m. *stands for* »*limit in the mean*») *will exist, and*

$$f(x) = \underset{A\to\infty}{\text{l.i.m.}} \frac{1}{\sqrt{2\pi}} \int_{-A}^{A} g(u) e^{iux}\, du. \tag{1.03}$$

$g(u)$ is known as the »Fourier transform» of $f(x)$. To prove this, let us put

$$f_A(x) = \begin{cases} f(x) & \text{if } |x| < A, \\ 0 & \text{if } |x| \geq A. \end{cases} \tag{1.04}$$

Let us represent $f_A(x)$ over $(-2A, 2A)$ by the Fourier series

$$f_A(x) \sim \sum_{-\infty}^{\infty} a_n e^{\frac{in\pi x}{2A}}. \tag{1.05}$$

Then

$$\int_{-\infty}^{\infty} f_A(x+\xi) \bar{f}_A(\xi)\, d\xi = \int_{-2A}^{2A} f_A(x+\xi) \bar{f}_A(\xi)\, d\xi$$

$$= \sum_{-\infty}^{\infty} a_n \bar{a}_n \int_{-2A}^{2A} e^{\frac{in\pi(x+\xi)}{2A}} e^{-\frac{in\pi\xi}{2A}}\, d\xi$$

$$= \sum_{-\infty}^{\infty} 4A\, |a_n|^2 e^{\frac{in\pi x}{2A}}. \tag{1.06}$$

This series of equations merits several comments. First, the infinite integrals which appear are infinite in appearance only, as the integrand vanishes beyond a certain point. Secondly, the period chosen for the Fourier representation of $f_A(x)$ is twice the length of the interval over which $f_A(x)$ may differ from 0, so that one period of $f_A(x+\xi)$ may overlap not more than one corresponding period of $\bar{f}_A(\xi)$. Third, the function $\int_{-\infty}^{\infty} f_A(x+\xi)\bar{f}_A(\xi)\, d\xi$ has a Fourier development which possesses only positive coefficients, and is absolutely and uniformly convergent, as follows at once from the Hurwitz theorem. The positiveness of the Fourier coefficients of this function forms the point of departure for the greater part of the present paper.

16—29764. *Acta mathematica.* 55. Imprimé le 7 avril 1930.

It follows at once that

$$\int_{-\infty}^{\infty} f_A(x+\xi) f_A(\xi)\, d\xi = \lim_{N\to\infty} \frac{1}{4A} \int_{-2A}^{2A} \frac{\sin \frac{2N+1}{4A}\pi(x-y)}{\sin \frac{\pi(x-y)}{4A}}\, dy \int_{-\infty}^{\infty} f_A(y+\xi) \bar{f}_A(\xi)\, d\xi. \quad (1.07)$$

However, Lebesgue's fundamental theorem on the Fourier coefficients, to the effect that they always tend to zero, yields us

$$\lim_{N\to\infty} \frac{1}{4A} \int_{-2A}^{2A} \sin \frac{2N+1}{4A}\pi(x-y) \left[\frac{1}{\frac{\pi(x-y)}{4A}} - \frac{1}{\sin \frac{\pi(x-y)}{4A}} \right] \cdot$$

$$\cdot\, dy \int_{-\infty}^{\infty} f_A(y+\xi) \bar{f}_A(\xi)\, d\xi = 0. \quad (1.08)$$

Combining these two relations, we see that

$$\int_{-A}^{A} |f(x)|^2 dx = \int_{-\infty}^{\infty} |f_A(x)|^2\, dx$$

$$= \lim_{N\to\infty} \frac{1}{\pi} \int_{-2A}^{2A} \frac{\sin \frac{2N+1}{4A}\pi y}{y}\, dy \int_{-\infty}^{\infty} f_A(y+\xi) \bar{f}_A(\xi)\, d\xi$$

$$= \lim_{N\to\infty} \frac{1}{\pi} \int_{-\infty}^{\infty} dy \int_0^{\frac{2N+1}{4A}\pi} \cos uy\, du \int_{-\infty}^{\infty} f_A(y+\xi) \bar{f}_A(\xi)\, d\xi$$

$$= \frac{1}{2\pi} \int_{-\infty}^{\infty} du \int_{-\infty}^{\infty} f_A(\eta)\, d\eta \int_{-\infty}^{\infty} \bar{f}_A(\xi)\, e^{iu(\eta-\xi)}\, d\xi$$

$$= \frac{1}{2\pi} \int_{-\infty}^{\infty} du \left| \int_{-A}^{A} f(\eta)\, e^{iu\eta}\, d\eta \right|^2. \quad (1.09)$$

The inversions of the order of integration are here justified by the fact that all the infinite limits are merely apparent, and are introduced to simplify the formal work of inversion. If we replace $f_A(x)$ by the function $f_B(x)-f_A(x)$ which has essentially the same properties, we see that

$$\int_{-B}^{B} |f(x)|^2\,dx - \int_{-A}^{A} |f(x)|^2\,dx = \frac{1}{2\pi}\int_{-\infty}^{\infty} du \left| \int_{-B}^{B} f(\eta)\,e^{iu\eta}\,d\eta - \int_{-A}^{A} f(\eta)e^{iu\eta}\,d\eta \right|^2. \tag{1.10}$$

In case $\int_{-\infty}^{\infty} |f(x)|^2\,dx$ exists,

$$\lim_{B,\,A\to\infty} \int_{-\infty}^{\infty} du \left| \int_{-B}^{B} f(\eta)\,e^{iu\eta}\,d\eta - \int_{-A}^{A} f(\eta)\,e^{iu\eta}\,d\eta \right|^2 = 0 \tag{1.11}$$

and we may use Weyl's lemma to the Riesz-Fischer theorem to prove that

$$g(u) = \underset{A\to\infty}{\text{l.i.m.}} \frac{1}{\sqrt{2\pi}} \int_{-A}^{A} f(\eta)\,e^{iu\eta}\,d\eta \tag{1.12}$$

exists, and is »quadratically summable». Combining this definition of $g(u)$ with (1.09), we see that

$$\int_{-\infty}^{\infty} |g(u)|^2\,du = \int_{-\infty}^{\infty} |f(x)|^2\,dx. \tag{1.13}$$

That is, the integral of the square of the modulus of a function is invariant under a Fourier transformation.

To complete the proof of Plancherel's theorem, it is merely necessary to show that for functions $f(x)$ of some closed set,

$$f(x) = \underset{A\to\infty}{\text{l.i.m.}} \frac{1}{\sqrt{2\pi}} \int_{-A}^{A} g(u)\,e^{-iux}\,du. \tag{1.14}$$

A particular choice of $f(x)$ is the following:

$$f(x) = \begin{cases} 0; & [x < \alpha] \\ 1; & [\alpha < x < \beta] \\ 0; & [\beta < x]. \end{cases} \tag{1.15}$$

Here

$$g(u)=\frac{1}{\sqrt{2\pi}}\int_{\alpha}^{\beta} e^{iux}\,dx=\frac{e^{iu\beta}-e^{iu\alpha}}{iu\sqrt{2\pi}}. \tag{1. 16}$$

Hence

$$\underset{A\to\infty}{\mathrm{l.i.m.}}\frac{1}{\sqrt{2\pi}}\int_{-A}^{A} g(u)e^{-iux}\,du=\underset{A\to\infty}{\mathrm{l.i.m.}}\frac{1}{2\pi}\int_{-A}^{A}\frac{e^{iu(\beta-x)}-e^{iu(\alpha-x)}}{iu}\,du$$

$$=\underset{A\to\infty}{\mathrm{l.i.m.}}\frac{1}{2\pi}\int_{-A}^{A}\frac{\sin u(\beta-x)-\sin u(\alpha-x)}{u}\,du$$

$$=\frac{1}{2}\left[\operatorname{sgn}(\beta-x)-\operatorname{sgn}(\alpha-x)\right]$$

$$=f(x) \tag{1. 17}$$

except possibly at the two points α and β, a set of zero measure. This completes the proof of Plancherel's theorem.

Plancherel states this theorem somewhat differently. He essentially defines $g(u)$ as

$$g(u)=\frac{d}{du}\frac{1}{\sqrt{2\pi}}\int_{-\infty}^{\infty} d\eta\int_{0}^{u} f(\eta)\,e^{iu\eta}\,dv. \tag{1. 18}$$

If we retain our definition, it follows from an elementary use of the Schwarz inequality that

$$\int_{0}^{u} g(v)\,dv=\frac{1}{\sqrt{2\pi}}\int_{-\infty}^{\infty} d\eta\int_{0}^{u} f(\eta)\,e^{iv\eta}\,dv. \tag{1. 19}$$

To see this, les us reflect that

$$\left|\int_{0}^{u} g(v)\,dv-\frac{1}{\sqrt{2\pi}}\int_{-A}^{A} d\eta\int_{0}^{u} f(\eta)e^{iv\eta}\,dv\right|=\lim_{B\to\infty}\left|\frac{1}{\sqrt{2\pi}}\int_{0}^{u} dv\left[\int_{A}^{B}+\int_{-B}^{-A}\right]f(\eta)\,e^{iv\eta}\,d\eta\right|$$

$$\leq\lim_{B\to\infty}\left\{\frac{1}{2\pi}\int_{0}^{u} dv\int_{0}^{u}\left|\left[\int_{A}^{B}+\int_{-B}^{-A}\right]f(\eta)\,e^{iv\eta}\,d\eta\right|^{2}dv\right\}^{1/2}$$

$$\leq u \lim_{B\to\infty} \left\{ \frac{1}{2\pi} \int_{-\infty}^{\infty} \left| \left[\int_{A}^{B} + \int_{-B}^{-A} \right] f(\eta)\, e^{iu\eta}\, d\eta \right|^2 dv \right\}^{1/2}$$

$$= u \left\{ \left[\int_{A}^{\infty} + \int_{-\infty}^{-A} \right] |f(\eta)|^2\, d\eta \right\}^{1/2},$$

and since $\int_{-\infty}^{\infty} |f(\eta)|^2\, d\eta$ is finite, it follows that

$$\lim_{A\to\infty} \left| \int_{0}^{u} g(v)\, dv - \frac{1}{\sqrt{2\pi}} \int_{-A}^{A} d\eta \int_{0}^{u} f(\eta)\, e^{iv\eta}\, dv \right| = 0.$$

From this (1. 19) follows at once. Since a summable function is almost everywhere the derivative of its integral, the transition to Plancherel's form of the definition is immediate.

It follows at once from Plancherel's theorem that if $f_1(x)$ and $f_2(x)$ are quadratically summable,

$$F_1(u) = \underset{A\to\infty}{\mathrm{l.i.m.}} \frac{1}{\sqrt{2\pi}} \int_{-A}^{A} f_1(x)\, e^{iux}\, dx \tag{1. 20}$$

and

$$F_2(u) = \underset{A\to\infty}{\mathrm{l.i.m.}} \frac{1}{\sqrt{2\pi}} \int_{-A}^{A} f_2(x)\, e^{iux}\, dx \tag{1. 21}$$

exist, and that

$$\int_{-\infty}^{\infty} |F_1(u) \pm F_2(u)|^2\, du = \int_{-\infty}^{\infty} |f_1(x) \pm f_2(x)|^2\, dx \tag{1. 22}$$

and

$$\int_{-\infty}^{\infty} |F_1(u) \pm iF_2(u)|^2\, du = \int_{-\infty}^{\infty} |f_1(x) \pm if_2(x)|^2\, dx. \tag{1. 23}$$

Combining the last four formulae with one another, we have

$$\int_{-\infty}^{\infty} F_1(u)\,\overline{F}_2(u)\,du = \int_{-\infty}^{\infty} f_1(x)\bar{f}_2(x)\,dx. \tag{1. 24}$$

This we may know as the Parseval theorem for the Fourier integral. Since

$$\overline{F}_2(-u) = \underset{A\to\infty}{\text{l.i.m.}}\,\frac{1}{\sqrt{2\pi}}\int_{-A}^{A} \bar{f}_2(x)\,e^{iux}\,dx \tag{1. 25}$$

we may deduce at once that

$$\int_{-\infty}^{\infty} F_1(u)\,F_2(-u)\,du = \int_{-\infty}^{\infty} f_1(x)f_2(x)\,dx. \tag{1. 26}$$

Since furthermore

$$\overline{F}_2(v-u) = \underset{A\to\infty}{\text{l.i.m.}}\,\frac{1}{\sqrt{2\pi}}\int_{-A}^{B} \bar{f}_2(x)\,e^{ivx}\,e^{iux}\,dx \tag{1. 27}$$

it follows that

$$\int_{-\infty}^{\infty} F_1(u)\,F_2(v-u)\,du = \int_{-\infty}^{\infty} f_1(x)f_2(x)\,e^{ivx}\,dx. \tag{1. 28}$$

As a consequence, if $f_1(x)f_2(x)$ is quadratically summable, its Fourier transform is

$$\sqrt{2\pi}\int_{-\infty}^{\infty} F_1(u)\,F_2(v-u)\,du. \tag{1. 29}$$

This theorem lies at the basis of the whole operational calculus.

2. Schuster's periodogram analysis.

The two theories of harmonic analysis embodied in the classical Fourier series development and the theory of Plancherel do not exhaust the possibilities of harmonic analysis. The Fourier series is restricted to the very special class of periodic functions, while the Plancherel theory is restricted to functions which are quadratically summable, and hence tend on the average to zero as their argument tends to infinity. Neither is adequate for the treatment of a ray of white light which is supposed to endure for an indefinite time. Nevertheless, the physicists who first were faced with the problem of analyzing white light into its components had to employ one or the other of these tools. Gouy accordingly represented white light by a Fourier series, the period of which he allowed to grow without limit, and by focussing his attention on the average values of the energies concerned, he was able to arrive at results in agreement with the experiments. Lord Rayleigh on the other hand, achieved much the same purpose by using the Fourier integral, and what we now should call Plancherel's theorem. In both cases one is astonished by the skill with which the authors use clumsy and unsuitable tools to obtain the right results, and one is led to admire the unfailing heuristic insight of the true physicist.

The net outcome of the work of these writers was to dispel the idea that white light consist in some physical, supermathematical way of homogeneous monochromatic vibrations. Schuster in particular, was led to the conclusion that when white light is analyzed by a grating, the monochromatic components are created by the grating rather than selected by it. Thus a great stimulus was given to the investigation of the sense in which any phenomenon may be said to contain hidden periodic components. The successful completion of this investigation is also due to Schuster.

Schuster sums up his conclusions as follows[2]: »Let y be a function of t, such that its values are regulated by some law of probability, not necessarily the exponential one, but acting in such a manner that if a large number of t be chosen at random, there will always be a definite fraction of that number depending on t_1 only, which lie between t_1 and t_1+T, where T is any given time interval.

»Writing

$$A=\int_{t_1}^{t_1+T} y\cos \varkappa t\,dt \quad \text{and} \quad B=\int_{t_1}^{t_1+T} y\sin \varkappa t\,dt,$$

and forming

$$R=\sqrt{A^2+B^2},$$

the quantity R will, with increasing values of T, fluctuate about some mean value, which increases proportionally to $\sqrt{T}$ provided T is taken sufficiently large.

»If this theorem is taken in conjunction with the two following well-known propositions,

(1) If $y = \cos \varkappa t$, R will, apart from periodical terms, increase proportionally to T;

(2) If $y = \cos \lambda t$, λ being different from $\varkappa$, the quantity R will fluctuate about a constant value;

it is seen that we have means at our disposal to separate any true periodicity of a variable from among its irregular changes, provided we can extend the time limits sufficiently. . . . The application of the theory of probability to the investigation of what may be called »hidden» periodicities . . . may be further extended . . .»

While Schuster's statement is perhaps not in all respects clear, it contains the germs of all subsequent generalizations of harmonic analysis. First among these is the emphasis on the notion of the *mean*. The operator which yields

$$A_1 = \lim_{T\to\infty} \frac{1}{T} \int_{t_1}^{t_1+T} y \cos \varkappa t \, dt \text{ or } B_1 = \lim_{T\to\infty} \frac{1}{T} \int_{t_1}^{t_1+T} y \sin \varkappa t \, dt \tag{2. 01}$$

annihilates all functions $y(t)$ made up in a purely fortuitous or haphazard manner, as well as all trigonometrical functions other than $\cos \varkappa t$ or $\sin \varkappa t$, respectively. Hence we may take A_1 and B_1 to indicate the amounts of $\cos \varkappa t$ or $\sin \varkappa t$ contained in y. As a simultaneous indication of these two quantities, neglecting phase, Schuster takes $\sqrt{A_1^2 + B_1^2}$ which he supersedes in his later papers by the somewhat simpler expression $A_1^2 + B_1^2$.

It is possible to lend a certain plausibility to this later choice of Schuster as contrasted with his earlier, by considering the expression

$$\varphi(x) = \lim_{T\to\infty} \frac{1}{2T} \int_{-T}^{T} f(x+t)\bar{f}(t)\, dt. \tag{2. 02}$$

If

$$f(t) = \sum_{1}^{N} a_n e^{i\lambda_n t}, \tag{2. 03}$$

we have

$$\varphi(x) = \sum_{1}^{N} |a_n|^2 e^{i\lambda_n x}. \tag{2. 04}$$

Accordingly,

$$\varrho(\lambda_n) = |a_n|^2 = \lim_{T\to\infty} \frac{1}{2T} \int_{-T}^{T} \varphi(x)\, e^{-i\lambda_n x}\, dx. \tag{2.05}$$

This function $\varphi(t)$ differs from $f(t)$ in that every amplitude of a trigonometric term in $f(t)$ is replaced by the square of its modulus.

The expression $|a_n|^2$ is necessarily positive. It is, however, unobservable in any actual case, as we only have a finite interval of time at our disposal. Let it be noted that if we put

$$\varphi'_T(x) = \frac{1}{2T} \int_{-T}^{T} f(x+t)\bar{f}(t)\, dt \tag{2.06}$$

and

$$\varrho'_T(\lambda_n) = \frac{1}{2T} \int_{-T}^{T} \varphi'_T(x)\, e^{-i\lambda_n x}\, dx, \tag{2.07}$$

it is *not* necessarily true that ϱ'_T is non-negative. On the other hand, if we put

$$f_A(t) = f(t)\ [|t| < A];\ f_A(t) = 0 \text{ otherwise} \tag{2.08}$$

and

$$\varphi_A(x) = \frac{1}{2A} \int_{-\infty}^{\infty} f_A(x+t)\bar{f}_A(t)\, dt \tag{2.09}$$

then

$$\varrho_A(\lambda_n) = \frac{1}{2A} \int_{-\infty}^{\infty} \varphi_A(x)\, e^{-i\lambda_n x}\, dx = \frac{1}{4A^2} \left| \int_{-\infty}^{\infty} f_A(t)\, e^{-i\lambda_n t}\, dt \right|^2 \geq 0. \tag{2.10}$$

This suggests an improved method of treating the approximate periodogram of a function under observation for a finite time.

The periodogram of a function — that is, the graph of the discontinuous function $\varrho(\lambda_n)$ or its approximate continuous analyses $\varrho_A(\lambda_n)$ — contains but a small amount of the information which the complete graph of the original function is able to yield. Not only do we deliberately discard all phase relations, but a large part of the original function — often the most interesting and important part — is thrown away as the aperiodic residue. The chief reason for this that any measure for a continuous spectral density becomes infinite at a spectral line, while any measure for the intensity of a spectral line becomes zero over the continuous spectrum.

This is a difficulty, however, which has had to be faced in many other branches of mathematics and physics. Impulses and forces are treated side by side in mechanics, although they have no common unit. We are familiar in potential theory with distributions of charge containing point, line, and surface distributions, as well as continuous volume distributions. The basic theory of all these problems is that of the Stieltjes integral.

Let us put

$$S(u) = \frac{1}{2\pi}\int_{-\infty}^{\infty} \varphi(x)\frac{e^{ixu}-1}{ix}\,dx. \qquad (2.11)$$

Here the term 1 is introduced to cancel the singularity which we should otherwise find for $x=0$. We have formally and heuristically

$$\begin{aligned} S(u+0)-S(u-0) &= \lim_{\varepsilon\to 0}\int_{-\infty}^{\infty}\varphi(x)\,\frac{e^{ix(u+\varepsilon)}-e^{ix(u-\varepsilon)}}{2\pi i x}\,dx \\ &= \lim_{\varepsilon\to 0}\frac{1}{\pi}\int_{-\infty}^{\infty}\varphi(x)\,e^{iux}\,\frac{\sin\varepsilon x}{x}\,dx \\ &= \lim_{\eta\to 0}\frac{1}{\pi\eta}\int_{0}^{\eta}d\varepsilon\int_{-\infty}^{\infty}\varphi(x)\,e^{iux}\,\frac{\sin\varepsilon x}{x}\,dx \\ &= \lim_{\eta\to 0}\frac{\eta}{\pi}\int_{-\infty}^{\infty}\varphi(x)\,e^{iux}\,\frac{1-\cos\eta x}{\eta^2 x^2}\,dx \\ &= \lim_{T\to\infty}\frac{1}{\pi T}\int_{-\infty}^{\infty}\varphi(x)\,e^{iux}\,\frac{T^2\left(1-\cos\frac{x}{T}\right)}{x^2}\,dx. \end{aligned} \qquad (2.12)$$

Now,

$$\frac{T^2\left(1-\cos\frac{x}{T}\right)}{x^2} \qquad (2.13)$$

is a positive function assuming the value $1/2$ for $x=0$, with a graph with a scale in the x direction proportional to T, and with a finite integral. Hence, it does not seem amiss to consider $S(u+0)-S(u-0)$ except for a constant factor, as the same expression as $\varrho(u)$. We shall later verify this fact in more detail and with more rigor. On the other hand, again formally,

$$S'(u) = \frac{1}{2\pi}\int_{-\infty}^{\infty} \varphi(x)\, e^{ixu}\, dx. \tag{2. 14}$$

Thus in case $\varphi(x)$ is of too small an order of magnitude to possess a line spectrum, $S(u)$ still has a significance. We shall interpret its derivative as meaning the density of the continuous portion of the spectrum of $f(t)$.

The graph of $S(u)$ shall be called the *integrated periodogram* of $f(t)$. We shall show later that under very general conditions, it may be so chosen as to be a monotone non-decreasing curve. The amount of rise of this curve between the arguments indicates the total intensity of the part of the spectrum lying between the frequencies. This shift of our attention from the periodogram itself to the integrated periodogram, which is monotone but not necessarily everywhere differentiable, is as we have said of the same nature as the shift from $g(x)$ in

$$\int f(x)\, g(x)\, dx \tag{2. 15}$$

to $\alpha(x)$ in

$$\int f(x)\, d\alpha(x). \tag{2. 16}$$

I wish to remark in passing that the formulae for the integrated periodogram are at least as convenient for computational purposes as the formulae of the Schuster analysis, that the monotony of the intergrated periodogram avoids the possibility of overlooking important periods by an insufficient search, while it gives an immediate indication of empty parts of the spectrum which need no further exploration; and that the computation of $\varphi(x)$ and $S(u)$ may be performed by such instruments as the product integraph of V. Bush. I also wish to call attention to a practical study of these modified periodogram methods by Mr. G. W. Kenrick of the Massachusetts Institute of Technology.

CHAPTER II.

3. The spectrum of an arbitrary function of a single variable.

The present section is devoted to the rigorous delimitation and demonstration of the theorems heuristically indicated in section 2. Let $f(t)$ be a measurable function such that

$$\varphi(x)=\lim_{T\to\infty}\frac{1}{2T}\int_{-T}^{T} f(x+t)\bar{f}(t)\,dt \tag{3.01}$$

exists for every x. This is the sole assumption necessary in the present section. By the Schwarz inequality

$$|\varphi(x)|\leq \lim_{T\to\infty}\frac{1}{2T}\left\{\int_{-T}^{T}|f(x+t)|^2\,dt\int_{-T}^{T}|f(t)|^2\,dt\right\}^{1/2}. \tag{3.02}$$

It follows from this that $\varphi(x)$ is bounded. To show this, it is only necessary to prove that

$$\lim_{T\to\infty}\frac{1}{2T}\int_{-T}^{T}|f(x+t)|^2\,dt=\lim_{T\to\infty}\frac{1}{2T}\int_{-T}^{T}|f(t)|^2\,dt=\varphi(0). \tag{3.03}$$

We have

$$\left|\frac{1}{2T}\int_{-T}^{T}|f(x+t)|^2\,dt-\frac{1}{2T}\int_{-T}^{T}|f(t)|^2\,dt\right|$$

$$=\left|\frac{1}{2T}\int_{T}^{T+x}|f(t)|^2\,dt-\frac{1}{2T}\int_{-T}^{-T+x}|f(t)|^2\,dt\right|$$

$$\leq\left|\frac{1}{2T}\int_{T}^{T+x}|f(t)|^2\,dt+\frac{1}{2T}\int_{-T}^{-T+x}|f(t)|^2\,dt\right|$$

$$\leq\left|\frac{1}{2T}\int_{-T-x}^{T+x}|f(t)|^2\,dt-\frac{1}{2T}\int_{-T+x}^{T-x}|f(t)|^2\,dt\right|$$

$$=\left|\left(1+\frac{x}{T}\right)\frac{1}{2(T+x)}\int_{-T-x}^{T+x}|f(t)|^2\,dt-\left(1-\frac{x}{T}\right)\frac{1}{2(T-x)}\int_{-T+x}^{T-x}|f(t)|^2\,dt\right|. \tag{3.04}$$

Hence

$$\overline{\lim_{T\to\infty}} \left| \frac{1}{2T}\int_{-T}^{T} |f(x+t)|^2\,dt - \frac{1}{2T}\int_{-T}^{T} |f(t)|^2\,dt \right|$$

$$\leq \lim_{T\to\infty} \left| \left(1+\frac{x}{T}\right)\frac{1}{2(T+x)}\int_{-T-x}^{T+x} |f(t)|^2\,dt - \left(1-\frac{x}{T}\right)\frac{1}{2(T-x)}\int_{-T+x}^{T-x} |f(t)|^2\,dt \right| = 0. \tag{3.05}$$

Therefore

$$|\varphi(x)| \leq \varphi(0), \tag{3.06}$$

and $\varphi(x)$ is bounded.

As before, we put

$$\varphi_A(x) = \frac{1}{2A}\int_{-\infty}^{\infty} f_A(x+t)\bar{f}_A(t)\,dt. \tag{3.07}$$

By the Schwarz inequality

$$|\varphi_A(x)| \leq \frac{1}{2A}\sqrt{\int_{-\infty}^{\infty} |f_A(x+t)|^2\,dt \int_{-\infty}^{\infty} |f_A(t)|^2\,dt}$$

$$= \frac{1}{2A}\int_{-\infty}^{\infty} |f_A(t)|^2\,dt, \tag{3.08}$$

and $\varphi_A(x)$ is uniformly bounded in x and A for all values of A larger than some given value. Furthermore, if $x > 0$,

$$\varphi_A(x) = \frac{1}{2A}\int_{-A}^{A-x} f(x+t)\bar{f}(t)\,dt$$

$$= \frac{1}{2A}\int_{-A}^{A} f(x+t)\bar{f}(t)\,dt - \frac{1}{2A}\int_{A-x}^{A} f(x+t)\bar{f}(t)\,dt. \tag{3.09}$$

We shall have a similar formula in case x is negative. We have furthermore

$$\left|\frac{1}{2A}\int_{A-x}^{A} f(x+t)f(t)\,dt\right| \leq \frac{1}{2A}\sqrt{\int_{A}^{A+x}|f(t)|^2\,dt\int_{A-x}^{A}|f(t)|^2\,dt}$$

$$\leq \frac{1}{2A}\left|\int_{A-x}^{A+x}|f(t)|^2\,dt\right|. \tag{3.10}$$

Since

$$\lim_{A\to\infty}\frac{1}{2A}\left|\int_{A-x}^{A+x}|f(t)|^2\,dt\right| = 0, \tag{3.11}$$

it follows at once that

$$\lim_{A\to\infty}\varphi_A(x) = \lim_{A\to\infty}\frac{1}{2A}\int_{-A}^{A} f(x+t)\bar{f}(t)\,dt = \varphi(x). \tag{3.12}$$

Thus $\varphi(x)$ is the limit of a uniformly bounded sequence of measurable functions, and is measurable. Since it is also bounded, it is quadratically summable over any finite range, while $\varphi(x)/x$ is quadratically summable over any range excluding the origin. It is, moreover, easy to prove that

$$\varphi(x) = \underset{A\to\infty}{\text{l.i.m.}}\ \varphi_A(x) \tag{3.13}$$

over any finite range, and that

$$\frac{\varphi(x)}{x} = \underset{A\to\infty}{\text{l.i.m.}}\ \frac{\varphi_A(x)}{x} \tag{3.14}$$

over any range excluding the origin. Hence,

$$\varphi(x)\frac{\sin\mu x}{x} = \underset{A\to\infty}{\text{l.i.m.}}\ \varphi_A(x)\frac{\sin\mu x}{x}. \tag{3.15}$$

In as much as the Fourier transformation leaves invariant the integral of the square of the modulus of a function, and hence leaves invariant all properties of convergence in the mean,

$$\underset{N\to\infty}{\text{l.i.m.}}\int_{-N}^{N}\varphi(x)\frac{\sin\mu x}{x}e^{iux}\,dx = \underset{A\to\infty}{\text{l.i.m.}}\int_{-\infty}^{\infty}\varphi_A(x)\frac{\sin\mu x}{x}e^{iux}\,dx$$

$$= \underset{A\to\infty}{\text{l.i.m.}} \int_{-\infty}^{\infty} \varphi_A(x)\, e^{iux}\, dx \int_0^{\mu} \cos \xi x\, d\xi$$

$$= \frac{1}{2} \underset{A\to\infty}{\text{l.i.m.}} \int_0^{\mu} d\xi \int_{-\infty}^{\infty} \varphi_A(x) [e^{i(u+\xi)x} + e^{i(u-\xi)x}]\, dx. \qquad (3.16)$$

The inversion of the order of integration is justified as usual by the fact that the infinite integral is only apparently infinite. This, let me remark parenthetically, is the case also in the next set of formulae.

The last expression is the limit in the mean of a real non-negative quantity, for

$$\int_{-\infty}^{\infty} \varphi_A(x) e^{ivx} dx = \frac{1}{2A} \int_{-\infty}^{\infty} e^{ivx}\, dx \int_{-\infty}^{\infty} f_A(x+t) \bar{f}_A(t)\, dt$$

$$= \frac{1}{2A} \int_{-\infty}^{\infty} \bar{f}_A(t)\, dt \int_{-\infty}^{\infty} f_A(x+t)\, e^{ivx}\, dx$$

$$= \frac{1}{2A} \int_{-\infty}^{\infty} \bar{f}_A(t)\, dt \int_{-\infty}^{\infty} f_A(w)\, e^{iv(w-t)}\, dw$$

$$= \frac{1}{2A} \left| \int_{-\infty}^{\infty} f_A(w)\, e^{ivw}\, dw \right|^2 \geq 0. \qquad (3.17)$$

The limit in the mean of a function is determined with the exception of a set of points of zero measure, but the limit in the mean of a non-negative function may always be so chosen as to be non-negative. If we make this choice,

$$\underset{N\to\infty}{\text{l.i.m.}} \int_{-N}^{N} \varphi(x) \frac{\sin \mu x}{x} e^{iux}\, dx \geq 0. \qquad (3.18)$$

The expression

$$\sigma_1(u) = \underset{A\to\infty}{\text{l.i.m.}} \frac{1}{2\pi} \left[\int_1^A + \int_{-A}^{-1} \right] \varphi(x) \frac{e^{ixu}}{ix}\, dx \qquad (3.19)$$

exists, as the Fourier transform of a quadratically summable function. Moreover,

$$\sigma_2(u) = \frac{1}{2\pi}\int_{-1}^{1} \varphi(x)\frac{e^{ixu}-1}{ix}\,dx \tag{3.20}$$

exists. If we put

$$\sigma(u) = \sigma_1(u) + \sigma_2(u) \tag{3.21}$$

we have

$$\sigma(u+\mu)-\sigma(u-\mu) = \underset{A\to\infty}{\mathrm{l.i.m.}}\ \frac{1}{\pi}\int_{-A}^{A} \varphi(x)\frac{\sin \mu x}{x} e^{iux}\,dx \geq 0. \tag{3.22}$$

Of course, when we say that a limit in the mean is non-negative, we merely mean that it can be so chosen. Thus the expression $\sigma(u)$ is monotone, or at least can be so chosen, for example, by putting

$$\sigma(u) = \frac{d}{du}\int_0^u \sigma(u)\,du \tag{3.23}$$

at every point where the latter expression is defined. Here we introduce (3.23), because $\sigma(u)$ is now almost everywhere the limit of the difference quotient of $\int_0^u \sigma(u)\,du$, namely, $\frac{1}{2\varepsilon}\int_{u-\varepsilon}^{u+\varepsilon} \sigma(u)\,du$, which is monotone as a consequence of (3.22). Thus, except at a set of zero measure, $\sigma(u)$ is the limit, not merely the limit in the mean, of a set of monotone functions, and is monotone. Elsewhere, at a set of zero measure, we put

$$\sigma(u) = \frac{1}{2}\left[\sigma(u+0)+\sigma(u-0)\right]. \tag{3.24}$$

It follows that $\sigma(u+\mu)-\sigma(u-\mu)$ is of limited total variation over any finite interval. We shall show in the next paragraph that

$$\lim_{\mu\to\infty}\left[\sigma(u+\mu)-\sigma(u-\mu)\right]$$

is finite, and that hence $\sigma(u+\mu)-\sigma(u-\mu)$ is of limited total variation over $(-\infty, \infty)$. It is moreover, quadratically summable, as the Fourier transform of a quadratically summable function. It tends to o as $u \to \pm\infty$ and hence, by a theorem of Hobson[3], we have

$$\frac{1}{2}[\sigma(u+0+\mu)-\sigma(u+0-\mu)+\sigma(u-0+\mu)-\sigma(u-0-\mu)]$$

$$= \frac{1}{\pi}\int_{-\infty}^{\infty} \varphi(x)\frac{\sin \mu x}{x} e^{iux}\,dx. \qquad (3.25)$$

In particular, if $u=\mu=v/2$,

$$\frac{1}{2}[\sigma(v+0)+\sigma(v-0)] - \frac{1}{2}[\sigma(+0)+\sigma(-0)] = \frac{1}{2\pi}\int_{-\infty}^{\infty} \varphi(x)\frac{e^{ivx}-1}{ix}\,dx. \qquad (3.26)$$

If therefore we define

$$S(u) = \frac{1}{2\pi}\int_{-\infty}^{\infty} \varphi(x)\frac{e^{iux}-1}{ix}\,dx, \qquad (3.27)$$

$S(u)$ will exist, and

$$S(u)-\sigma(u) = \text{constant}. \qquad (3.28)$$

4. The total spectral intensity.

It is manifest that $\lim_{\mu\to\infty}[S(u+\mu)-S(u-\mu)]$, or as we shall write it, $S(\infty)-$ $-S(-\infty)$, if it exists, is a measure of the total spectral intensity of $f(x)$. We shall prove that this quantity exists and is finite.

We have

$$\frac{1}{A}\int_0^A [\sigma(u+\mu)-\sigma(u-\mu)]\,d\mu = \frac{1}{\pi A}\int_0^A d\mu \operatorname*{l.i.m.}_{B\to\infty}\int_{-B}^{B} \varphi(x)\frac{\sin \mu x}{x} e^{iux}\,dx. \qquad (4.01)$$

The limit in the mean is here taken with u as the fundamental variable, and with μ as parameter. It is not difficult to deduce from the boundedness of

$$\int_0^A d\mu \int_{-\infty}^{\infty} |\varphi(x)|^2 \frac{\sin^2 \mu x}{x^2}\,dx$$

that we may invert the order of integration, and get

18—29764. *Acta mathematica.* 55. Imprimé le 7 avril 1930.

$$\frac{1}{A}\int_0^A [\sigma(u+\mu)-\sigma(u-\mu)]\,d\mu = \frac{1}{\pi A}\operatorname*{l.i.m.}_{B\to\infty}\int_{-B}^{B}\varphi(x)\frac{1-\cos Ax}{x^2}e^{iux}\,dx. \tag{4.02}$$

To show this, let us remark that

$$\int_{-\infty}^{\infty}\left|\frac{1}{\pi A}\int_0^A d\mu \operatorname*{l.i.m.}_{B\to\infty}\int_{-B}^{B}\varphi(x)\frac{\sin\mu x}{x}e^{iux}\,dx - \frac{1}{\pi A}\int_0^A d\mu\int_{-C}^{C}\varphi(x)\frac{\sin\mu x}{x}e^{iux}\,dx\right|^2 du$$

$$= \frac{1}{\pi^2 A^2}\int_{-\infty}^{\infty}\left|\int_0^A d\mu \operatorname*{l.i.m.}_{B\to\infty}\left[\int_C^B + \int_{-B}^{-C}\right]\varphi(x)\frac{\sin\mu x}{x}e^{iux}\,dx\right|^2 du$$

$$\leq \frac{1}{\pi^2 A^2}\int_{-\infty}^{\infty} du\int_0^A d\mu_1\int_0^A\left|\operatorname*{l.i.m.}_{B\to\infty}\left[\int_C^B + \int_{-B}^{-C}\right]\varphi(x)\frac{\sin\mu x}{x}e^{iux}\,dx\right|^2 d\mu$$

$$= \frac{1}{\pi^2 A}\int_0^A d\mu\int_{-\infty}^{\infty} du\left|\operatorname*{l.i.m.}_{B\to\infty}\left[\int_C^B + \int_{-B}^{-C}\right]\varphi(x)\frac{\sin\mu x}{x}e^{iux}\,dx\right|^2$$

$$= \frac{2}{\pi A}\int_0^A d\mu\left[\int_C^{\infty} + \int_{-\infty}^{-C}\right]|\varphi(x)|^2\frac{\sin^2\mu x}{x^2}\,dx. \tag{4.03}$$

Inasmuch as this latter expression tends to 0 with increasing C,

$$\frac{1}{\pi A}\int_0^A d\mu \operatorname*{l.i.m.}_{B\to\infty}\int_{-B}^{B}\varphi(x)\frac{\sin\mu x}{x}e^{iux}\,dx = \operatorname*{l.i.m.}_{C\to\infty}\frac{1}{\pi A}\int_0^A d\mu\int_{-C}^{C}\varphi(x)\frac{\sin\mu x}{x}e^{iux}\,dx$$

$$= \operatorname*{l.i.m.}_{B\to\infty}\frac{1}{\pi A}\int_{-B}^{B}\varphi(x)\frac{e^{iux}}{x}\,dx\int_0^A \sin\mu x\,d\mu$$

$$= \frac{1}{\pi A}\operatorname*{l.i.m.}_{B\to\infty}\int_{-B}^{B}\varphi(x)\frac{1-\cos Ax}{x^2}e^{iux}\,dx, \tag{4.04}$$

thus proving our statement.

Our limit in the mean may be replaced by an ordinary limit, as this limit exists, owing to the boundedness of $\varphi(x)$. Therefore

$$\lim_{A\to\infty} \frac{1}{A} \int_0^A [\sigma(u+\mu)-\sigma(u-\mu)]\, d\mu = \lim_{A\to\infty} \frac{1}{\pi A} \int_{-\infty}^{\infty} \varphi(x) \frac{1-\cos Ax}{x^2} e^{iux}\, dx. \qquad (4.05)$$

It follows from the monotony of $\sigma(u+\mu)-\sigma(u-\mu)$ in μ that we may write

$$\begin{aligned}\lim_{\mu\to\infty} [\sigma(u+\mu)-\sigma(u-\mu)] &= \lim_{A\to\infty} \frac{1}{\pi} \int_{-\infty}^{\infty} \varphi\left(\frac{x}{A}\right) e^{iu\frac{x}{A}} \frac{1-\cos x}{x^2}\, dx \\ &\leq \max \left|\varphi\left(\frac{x}{A}\right)\right| \frac{1}{\pi} \int_{-\infty}^{\infty} \frac{1-\cos x}{x^2}\, dx \\ &= \varphi(0). \end{aligned} \qquad (4.06)$$

This yields us the existence of $\sigma(\infty)-\sigma(-\infty)$ and hence, according to the last paragraph, of $S(\infty)-S(-\infty)$. We have

$$\begin{aligned} S(\infty)-S(-\infty) &= \lim_{A\to\infty} \frac{1}{\pi} \int_{-\infty}^{\infty} \varphi\left(\frac{x}{A}\right) \frac{1-\cos x}{x^2}\, dx \\ &= \varphi(0) + \lim_{A\to\infty} \frac{1}{\pi} \int_{-\infty}^{\infty} \left[\varphi\left(\frac{x}{A}\right) - \varphi(0)\right] \frac{1-\cos x}{x^2}\, dx. \end{aligned} \qquad (4.07)$$

Hence for sufficiently large A

$$\begin{aligned} &|S(\infty)-S(-\infty)-\varphi(0)| \\ &\leq \frac{2}{\pi} \max|\varphi(\xi)| \left[\int_{A^{1/2}}^{\infty} + \int_{-\infty}^{-A^{1/2}}\right] \frac{1-\cos x}{x^2}\, dx + \max_{|\xi|<A^{-1/2}} |\varphi(\xi)-\varphi(0)| + \varepsilon. \end{aligned} \qquad (4.08)$$

Since A is arbitrary,

$$|S(\infty)-S(-\infty)-\varphi(0)| \leq \overline{\lim_{|\xi|\to 0}} |\varphi(\xi)-\varphi(0)|. \qquad (4.09)$$

In case $\varphi(x)$ is continuous at the origin,

$$\varphi(0) = S(\infty)-S(-\infty). \qquad (4.10)$$

However, $\varphi(x)$ need not be continuous at the origin, even if $f(t)$ is everywhere continuous. Thus let $f(t)=\sin t^2$.

Then

$$\varphi(0)=\lim_{T\to\infty}\frac{1}{2T}\int_{-T}^{T}\sin^2 t^2\,dt$$

$$=\lim_{T\to\infty}\frac{1}{T}\int_{0}^{T}\frac{1-\cos 2t^2}{2}\,dt$$

$$=\frac{1}{2}-\lim_{T\to\infty}\frac{\sqrt{2}}{T}\int_{0}^{T\sqrt{2}}\cos u^2\,du$$

$$=\frac{1}{2}. \qquad (4.11)$$

since $\int_0^\infty \cos u^2\,du$ is a Fresnel integral, and equals $\frac{1}{2}\sqrt{\frac{\pi}{2}}$.

On the other hand, if $x \neq 0$, we have

$$\varphi(x)=\lim_{T\to\infty}\frac{1}{2T}\int_{-T}^{T}\sin(t+x)^2\sin t^2\,dt$$

$$=\lim_{T\to\infty}\frac{1}{4T}\int_{-T}^{T}[-\cos(2t^2+2tx+x^2)+\cos(2tx+x^2)]\,dt.$$

The second part of this mean obviously vanishes. Hence

$$\varphi(x)=-\lim_{T\to\infty}\frac{1}{4T}\int_{-T}^{T}\cos(2t^2+2tx+x^2)\,dt$$

$$=-\lim_{U\to\infty}\frac{1}{2U}\int_{0}^{U}\cos\left(u^2+\frac{x^2}{2}\right)du$$

$$= o\left[\cos\frac{x^2}{2}\int_0^\infty \cos u^2\,du - \sin\frac{x^2}{2}\int_0^\infty \sin u^2\,du\right]$$

$$= o. \tag{4.12}$$

since $\int_0^\infty \cos u^2\,du = \int_0^\infty \sin u^2\,du = \frac{1}{2}\sqrt{\frac{\pi}{2}}$.

Thus $\varphi(x)$ vanishes almost everywhere, $S(u)$ vanishes identically, and

$$\varphi(0) \neq S(\infty) - S(-\infty). \tag{4.13}$$

5. Tauberian theorems and spectral intensity.

In a recent paper, the author has proved the following general Tauberian theorem: *Let* $M_1(x)$ *and* $M_2(x)$ *be two functions bounded over every range* $(\varepsilon, 1/\varepsilon)$, *which are* $O\left(\frac{1}{x(\log x)^2}\right)$ *at* 0 *and* ∞. *Let* $M_1(x)$ *be measurable and non-negative, and let*

$$\int_0^\infty M_1(x)\,x^{iu}\,dx \neq 0. \qquad [-\infty < u < \infty] \tag{5.01}$$

Let $M_2(x)$ *be continuous, except for a finite number of finite jumps. Let* $f(x)$ *be a measurable function bounded below and such that*

(a) $$\lim_{\lambda\to 0\,[\infty]} \int_0^\infty f(\lambda x)\,M_1(x)\,dx = A\int_0^\infty M_1(x)\,dx;$$

(b) $$\int_0^\infty f(\lambda x)\,M_1(x)\,dx \text{ is bounded.} \qquad [0 < \lambda < \infty]$$

Then

$$\lim_{\lambda\to 0\,[\infty]} \int_0^\infty f(\lambda x)\,M_2(x)\,dx = A\int_0^\infty M_2(x)\,dx. \tag{5.02}$$

Here ∞ is put into brackets to indicate that at these points it may be consistently substituted for 0. There is manifestly no restriction in assuming

$f(x)$ non-negative, as the theorem, if true for a given $f(x)$, is unchanged as to its validity by the addition to $f(x)$ of a constant. The theorem assumes a more understandable form under the transformations

$$\left.\begin{array}{l} x = e^{\xi}; \quad \lambda = e^{-\eta}; \\ e^{\xi} M_1(e^{\xi}) = N_1(\xi); \quad e^{\xi} M_2(e^{\xi}) = N_2(\xi). \end{array}\right\} \tag{5.03}$$

It then becomes: *Let $N_1(\xi)$ and $N_2(\xi)$ be two bounded functions which are $O(\xi^{-2})$ at $\pm\infty$. Let $N_1(\xi)$ be measurable and non-negative, and let*

$$\int_{-\infty}^{\infty} N_1(\xi)\, e^{iu\xi}\, d\xi \neq 0. \qquad [-\infty < u < \infty] \tag{5.04}$$

Let $N_2(\xi)$ be continuous, except for a finite number of finite jumps. Let $g(\xi)$ be a non-negative measurable function such that

(a)
$$\lim_{\eta\to\infty} \int_{-\infty}^{\infty} g(\eta-\xi)\, N_1(\xi)\, d\xi = A \int_{-\infty}^{\infty} N_1(\xi)\, d\xi;$$

(b)
$$\int_{-\infty}^{\infty} g(\eta-\xi)\, N_1(\xi)\, d\xi \text{ is bounded.} \qquad [-\infty < \eta < \infty]$$

Then

(c)
$$\lim_{\eta\to\infty} \int_{-\infty}^{\infty} g(\eta-\xi)\, N_2(\xi)\, d\xi = A \int_{-\infty}^{\infty} N_2(\xi)\, d\xi. \tag{5.05}$$

The proof proceeds as follows: We shall symbolize by C the class of all functions $N_2(\xi)$, bounded and $O(\xi^{-2})$ at $\pm\infty$, and continuous except for a finite number of finite jumps, for which (c) is a consequence of (a) and (b) for all non-negative measurable functions $g(\xi)$. Among the functions in C are all functions $N_2(\xi)$ of the form

$$N_2(\xi) = \int_{-\infty}^{\infty} N_1(\eta)\, R(\eta-\xi)\, d\eta \tag{5.06}$$

for which $\int_{-\infty}^{\infty} |R(\eta)|\, d\eta$ converges, inasmuch as the double integral

$$\int_{-\infty}^{\infty} R(\zeta)\, d\zeta \int_{-\infty}^{\infty} g(\eta-\xi-\zeta)\, N_1(\xi)\, d\xi$$

is absolutely convergent, so that

$$\int_{-\infty}^{\infty} N_2(\xi)\, g(\eta-\xi)\, d\xi = \int_{-\infty}^{\infty} g(\eta-\xi)\, d\xi \int_{-\infty}^{\infty} N_1(\zeta+\xi)\, R(\zeta)\, d\zeta$$

$$= \int_{-\infty}^{\infty} R(\zeta)\, d\zeta \int_{-\infty}^{\infty} g(\eta-\xi-\zeta)\, N_1(\xi)\, d\xi, \tag{5.07}$$

and

$$\lim_{\eta\to\infty} \int_{-\infty}^{\infty} N_2(\xi)\, g(\eta-\xi)\, d\xi = \lim_{\eta\to\infty} \int_{-\infty}^{\infty} R(\zeta)\, d\zeta \int_{-\infty}^{\infty} g(\eta-\xi-\zeta)\, N_1(\xi)\, d\xi$$

$$= A \int_{-\infty}^{\infty} R(\zeta)\, d\zeta \int_{-\infty}^{\infty} N_1(\xi)\, d\xi = A \int_{-\infty}^{\infty} N_2(\xi)\, d\xi. \tag{5.08}$$

A particular example of such a function is furnished by

$$N_2(\xi) = \int_{-B}^{B} \nu_2(u)\, e^{-iu\xi}\, du \tag{5.09}$$

where $\nu_2(u)$ is continuous over $(-B, B)$, while its first derivative is continuous except for a finite number of discontinuities of the first kind, and

$$\nu_2(B) = \nu_2(-B) = 0. \tag{5.10}$$

To prove this, let us reflect that $N_1(\xi)$ and $N_2(\xi)$ are quadratically summable by assumption, and that

$$\nu_1(u) = \int_{-\infty}^{\infty} N_1(\xi)\, e^{iu\xi}\, d\xi \tag{5.11}$$

exists, as well as

$$\nu'_1(u) = \underset{E\to\infty}{\mathrm{l.i.m.}} \int_{-E}^{E} i\xi N_1(\xi)\, e^{iu\xi}\, d\xi. \tag{5.12}$$

By our hypothesis (5.04)

$$\nu_1(u) \neq 0. \qquad [-\infty < u < \infty] \tag{5.13}$$

Let us put

$$\mu(u) = \nu_2(u)/\nu_1(u). \tag{5.14}$$

Inasmuch as $\mu(u)$ is absolutely continuous, its derivative may be computed by the rules, and

$$\mu'(u) = \frac{\nu_1(u)\,\nu'_2(u) - \nu_2(u)\,\nu'_1(u)}{[\nu_1(u)]^2}. \tag{5.15}$$

I now say that we shall have

$$\begin{aligned} R(\zeta) &= \int_{-B}^{B} \mu(u)\, e^{iu\zeta}\, du \\ &= \frac{1}{i\zeta} \int_{-B}^{B} \mu'(u)\, e^{iu\zeta}\, du. \end{aligned} \tag{5.16}$$

Inasmuch as

$$\int_{-B}^{B} \mu'(u)\, e^{iu\zeta}\, du \tag{5.17}$$

is quadratically summable,

$$\int_{-\infty}^{\infty} |R(\zeta)|\, d\zeta \tag{5.18}$$

exists. Since the integrals involved converge absolutely,

$$\begin{aligned} \int_{-\infty}^{\infty} N_1(\eta)\, R(\eta-\xi)\, d\eta &= \int_{-\infty}^{\infty} N_1(\eta)\, d\eta \int_{-B}^{B} \mu(u)\, e^{iu(\eta-\xi)}\, du \\ &= \int_{-B}^{B} \mu(u)\, e^{-iu\xi}\, du \int_{-\infty}^{\infty} N_1(\eta) e^{iu\eta}\, d\eta \\ &= \int_{-B}^{B} \mu(u)\, \nu_1(u)\, e^{-iu\xi}\, du \\ &= \int_{-B}^{B} \nu_2(u)\, e^{-iu\xi}\, du = N_2(\xi). \end{aligned} \tag{5.19}$$

This justifies our evaluation of $R(\zeta)$, and proves that $N_2(\xi)$ belongs to C. The following are particular cases which may be probed to belong to C in this manner:

$$T_B(\xi) = \int_{-B}^{B} [e^{-|u|} - e^{-B}]\, e^{-i\xi u}\, du$$

$$= \int_{0}^{B} [e^{-u} - e^{-B}] \cos \xi u\, du$$

$$= \frac{1}{\xi} \int_{0}^{B} e^{-u} \sin \xi u\, du$$

$$= \frac{1 - e^{-B} \cos B\xi}{1 + \xi^2} - \frac{e^{-B} \sin B\xi}{\xi(1+\xi^2)}. \tag{5.20}$$

Again,

$$Q_B(\xi) = \frac{1}{\pi} \int_{-B}^{B} \left(1 - \frac{|u|}{B}\right) \frac{\sin u}{u} e^{-iu\xi}\, du$$

$$= \int_{B(\xi-1)}^{B(\xi+1)} \frac{1 - \cos z}{\pi z^2}\, dz$$

$$= \frac{1}{\pi B(\xi - 1)} - \frac{1}{\pi B(\xi + 1)} + O\left(\frac{1}{B^2 \xi^2}\right). \qquad [B\xi \to \pm \infty] \tag{5.21}$$

If we already know certain members of the class C, we may obtain new members of the class in the following manner: Let $V(\xi)$ be a function continuous, except for a finite number of finite jumps, such that, when any positive ξ is given, we can find two members of C, $V_1(\xi)$ and $V_2(\xi)$, such that

$$V_1(\xi) \leq V(\xi) \leq V_2(\xi), \tag{5.22}$$

while

$$\int_{-\infty}^{\infty} [V_2(\xi) - V_1(\xi)]\, d\xi < \varepsilon. \tag{5.23}$$

Then $V(\xi)$ itself belongs to C. For

19—29764. *Acta mathematica*. 55. Imprimé le 7 avril 1930.

$$\overline{\lim_{\eta\to\infty}}\left|\int_{-\infty}^{\infty} g(\eta-\xi)\, V(\xi)\, d\xi - A\int_{-\infty}^{\infty} V(\xi)\, d\xi\right|$$

$$\leq \overline{\lim_{\eta\to\infty}}\left|\int_{-\infty}^{\infty} g(\eta-\xi)\, V_1(\xi)\, d\xi - A\int_{-\infty}^{\infty} V_2(\xi)\, d\xi\right|$$

$$+ \overline{\lim_{\eta\to\infty}}\left|\int_{-\infty}^{\infty} g(\eta-\xi)\, V_2(\xi)\, d\xi - A\int_{-\infty}^{\infty} V_1(\xi)\, d\xi\right|$$

$$< 2A\varepsilon. \tag{5.24}$$

and since ε is arbitrarily small, this limit is 0. Furthermore, any linear combination of a finite number of members of C belongs to C.

As a particular case, we have

$$[1+(1+B)e^{-B}]^{-1}\, T_B(\xi) < \frac{1}{1+\xi^2} < [1-(1-B)\,e^{-B}]^{-1}\, T_B(\xi). \tag{5.25}$$

Inasmuch as

$$\lim_{B\to\infty}(1+B)\,e^{-B} = 0, \tag{5.26}$$

it follows at once that $\frac{1}{1+\xi^2}$ belongs to C. An exactly similar proof will show that the same thing holds of

$$\frac{p}{(\xi-q)^2-r^2}.$$

Again,

$$\lim_{A\to\infty} Q_A(\xi) = \operatorname{sgn}(\xi+1)-\operatorname{sgn}(\xi-1) = V(\xi), \tag{5.27}$$

and this convergence is uniform except in the neighborhood of ± 1 while we always have for $B > 0$

$$Q_A(\xi) < \frac{1}{B(\xi^2+1)} \text{ over } (1+\eta,\ \infty) \text{ and } (-\infty,\ -1-\eta) \quad [A \text{ large enough}] \tag{5.28}$$

Furthermore,

$$\int_{-\infty}^{\infty} Q_A(\xi)\, d\xi = 4. \tag{5.29}$$

Let us put

$$\left.\begin{aligned}(1+\eta)\,Q_A(\xi(1-\eta)) &= V_2(\xi);\\ (1-\eta)\,Q_A(\xi(1+\eta)) - \frac{1}{B(\xi^2+1)} &= V_1(\xi).\end{aligned}\right\} \tag{5.30}$$

We can so determine a large A when η and B are given, that for that A and all larger ones,

$$V_2(\xi) \geq V(\xi) \geq V_1(\xi). \tag{5.22}$$

We have

$$\int_{-\infty}^{\infty} [V_2(\xi) - V_1(\xi)]\,d\xi = 2\left[\frac{1+\eta}{1-\eta} - \frac{1-\eta}{1+\eta}\right] + \frac{\pi}{B} \tag{5.31}$$

which we may make arbitrarily small. Hence

$$\operatorname{sgn}(\xi+1) - \operatorname{sgn}(\xi-1)$$

belongs to C. As an immediate consequence, since

$$\operatorname{sgn}(\xi+\alpha) - \operatorname{sgn}(\xi+\beta)$$

may be shown by the same means to belong to C, any step function vanishing for large positive and negative arguments belongs to C, and hence any function continuous except for a finite set of discontinuities of the first kind, and vanishing outside of a finite interval, since the latter function may be penned in between two step functions enclosing an arbitrarily small area.

Now let $N_2(\xi)$ be a bounded function which is $O(\xi^{-2})$ at $\pm\infty$, and which is continuous except for a finite number of finite jumps. Let

$$|N_2(\xi)| < P/(\xi^2+1), \tag{5.32}$$

for all ξ. We put

$$V_1(\xi) = \begin{cases} N_2(\xi); & [|\xi| < M] \\ -P/(\xi^2+1); & [|\xi| \geq M] \end{cases} \tag{5.33}$$

$$V_2(\xi) = \begin{cases} N_2(\xi); & [|\xi| < M] \\ P/(\xi^2+1). & [|\xi| \geq M] \end{cases} \tag{5.34}$$

The functions $V_1(\xi)$ and $V_2(\xi)$ are sums of functions of C and functions of the form $\pm P/(\xi^2+1)$ which also belong to C. Hence, they themselves belong to C. We have

$$V_2(\xi) > N_2(\xi) > V_1(\xi) \tag{5.35}$$

and

$$\int_{-\infty}^{\infty} [V_2(\xi) - V_1(\xi)]\, d\xi < 2P[\pi - \tan^{-1} M] \tag{5.36}$$

which we may make as small as we like. Hence $N_2(\xi)$ belongs to C. This concludes the proof of our generalized Tauberian theorem.

As a corollary of our Tauberian theorem, Mr. S. B. Littauer has given a proof of the following theorem of Jacob: *If $f(t)$ is a measurable function, integrable in every finite interval of* $(0, \infty)$ *and if for some given* α $(0 \leq \alpha < 1)$

(a) $$\frac{1}{T^{1-\alpha}} \int_0^T |f(t)|\, dt < B \text{ for every } T;$$

(b) $$\lim_{T\to\infty} \frac{1}{T^{1-\alpha}} \int_0^T f(t)\, dt = A;$$

then

(c) $$\lim_{\varepsilon\to 0} \frac{\varepsilon^{1-\alpha}}{\gamma_\alpha \pi} \int_0^\infty f(t) \left(\frac{\sin \varepsilon t}{\varepsilon t}\right)^2 dt = A,$$

where

$$\gamma_\alpha = \frac{2^{\alpha-1}(1-\alpha)}{\Gamma(2+\alpha)\cos\frac{\pi\alpha}{2}}.$$

Furthermore, if $f(t)$ is measurable and non-negative (*or bounded below*), (c) *implies* (b).

The particular case of this theorem where $\alpha = 0$ had already been treated by Bochner, Hardy, and the present author.

In all theorems of this type, there is a close relation between the theorem which one obtains by letting λ become infinite and that which one obtains by letting λ become 0. This is to be explained by the fact that the general Tauberian theorem assumes a perfectly symmetric form when we make the substitutions

$$x = e^\xi;\ \lambda = e^\eta;\ M_1(x) = N_1(\xi)\, e^\xi;\ M_2(x) = N_2(\xi)\, e^\xi. \tag{5.03}$$

If we take

$$\left.\begin{aligned} &M_1(x)\,[\text{or } M_2(x)] = 1 \text{ if } 0 < x < 1;\ = 0 \text{ otherwise};\\ &M_2(x)\,[\text{or } M_1(x)] = \frac{4\sin^2 x/2}{x^2} \end{aligned}\right\} \tag{5.37}$$

in our general Tauberian theorem, since $\int_0^\infty M_1(x)\, e^{iux}\, dx \neq 0$ and $\int_0^\infty M_2(x)\, e^{iux}\, dx \neq 0$, we may deduce the conclusions of the theorem. We thus get the following result: *Let*

$$|f(x)| < B. \qquad [0 < x < \infty]$$

Then the two propositions,

(a) $$\lim_{T\to 0} \frac{1}{T} \int_0^T f(x)\, dx = A$$

and

(b) $$\lim_{T\to 0} \frac{2}{\pi} \int_0^\infty f(Tx) \frac{1-\cos x}{x^2}\, dx = A$$

are equivalent.

In the particular case where $f(x)$ is replaced by $\frac{1}{2}[\varphi(x)+\varphi(-x)]$, we see that

$$A = \lim_{r\to\infty} \frac{1}{\pi} \int_{-\infty}^{\infty} \varphi\left(\frac{x}{r}\right) \frac{1-\cos x}{x^2}\, dx \tag{5.38}$$

implies, and is implied by

$$A = \lim_{\varepsilon\to 0} \frac{1}{2\,\varepsilon} \int_{-\varepsilon}^{\varepsilon} \varphi(x)\, dx. \tag{5.39}$$

We have (see (3.27))

$$\int_{-\infty}^{\infty} e^{-i\lambda u}\, dS(u) = \lim_{\mu\to 0} \frac{1}{2\,\pi\mu} \int_{-\infty}^{\infty} e^{-i\lambda u}\, du \underset{A\to\infty}{\mathrm{l.i.m.}} \int_{-A}^{A} \varphi(x) \frac{\sin \mu x}{x} e^{iux}\, dx$$

$$= \lim_{\mu\to 0} \frac{1}{\mu\lambda} \varphi(\lambda) \sin \mu\lambda = \varphi(\lambda), \tag{5.40}$$

except possibly at a set of points of zero measure. To see this, it is only necessary to reflect that it follows from the definition of the Stieltjes integral that if $\alpha(x)$ is of limited total variation over $(-\infty, \infty)$,

$$\int_{-\infty}^{\infty} f(x)\, d\alpha(x) = \lim_{\varepsilon\to 0} \sum_{-\infty}^{\infty} f(u+2\,n\varepsilon)[\alpha(u+(2\,n+1)\varepsilon)-\alpha(u+(2\,n-1)\varepsilon)]$$

$$= \lim_{\varepsilon \to 0} \frac{1}{2\varepsilon} \int_{-\varepsilon}^{\varepsilon} du \sum_{-\infty}^{\infty} f(u + 2n\varepsilon)\,[\alpha(u+(2n+1)\varepsilon) - \alpha(u+(2n-1)\varepsilon)]$$

$$= \lim_{\varepsilon \to 0} \frac{1}{2\varepsilon} \int_{-\varepsilon}^{\varepsilon} f(u)\,[\alpha(u+\varepsilon) - \alpha(u-\varepsilon)]\,du. \tag{5.41}$$

Let us put

$$\Phi(\lambda) = \int_{-\infty}^{\infty} e^{-i\lambda u}\, d\,S(u). \tag{5.42}$$

This function will be defined for all real arguments, and we shall have

$$\Phi(\lambda+\varepsilon) - \Phi(\lambda) = -2i \int_{-\infty}^{\infty} e^{-iu\left(\lambda + \frac{\varepsilon}{2}\right)} \sin\frac{u\varepsilon}{2}\, d\,S(u). \tag{5.43}$$

Since the function $\sin u\varepsilon/2$ is uniformly bounded, and tends to 0 over every finite range of u as $\varepsilon \to 0$, while $e^{-iu\left(\lambda + \frac{\varepsilon}{2}\right)}$ has modulus 1, it follows that $\Phi(\lambda+\varepsilon) - \Phi(\lambda)$ is less than the sum of two terms, one of which is the total variation of $S(u)$ over a region receding towards infinity, while the other is less than the total variation of $S(u)$ multiplied by a factor tending to 0. Hence

$$\lim_{\varepsilon \to 0} [\Phi(\lambda+\varepsilon) - \Phi(\lambda)] = 0. \tag{5.44}$$

Thus the function $\Phi(\lambda)$ is continuous, and indeed, this proof shows it to be uniformly continuous. Hence

$$\lim_{\varepsilon \to 0} \frac{1}{2\varepsilon} \int_{x-\varepsilon}^{x+\varepsilon} \varphi(\xi)\,d\xi = \lim_{\varepsilon \to 0} \frac{1}{2\varepsilon} \int_{x-\varepsilon}^{x+\varepsilon} \Phi(\xi)\,d\xi = \Phi(x). \tag{5.45}$$

This gives another proof that

$$\lim_{\varepsilon \to 0} \frac{1}{2\varepsilon} \int_{-\varepsilon}^{\varepsilon} \varphi(\xi)\,d\xi = S(\infty) - S(-\infty), \tag{5.46}$$

and indeed proves considerably more.

It is thus possible to dispense with Tauberian theorems for this part of the theory. There is another point, however, where they play a more essential rôle. That is in the study of the generalized Fourier transform of a function.

Let $\lim_{T\to\infty}\frac{1}{2T}\int_{-T}^{T}|f(x)|^2\,dx$ exist. Then

$$\left[\int_1^T+\int_{-T}^{-1}\right]\frac{|f(x)|^2}{x^2}\,dx=\int_1^T\frac{|f(x)|^2+|f(-x)|^2}{x^2}\,dx$$

$$=\int_1^T\frac{1}{x^2}\,d\int_0^x[|f(\xi)|^2+|f(-\xi)|^2]\,d\xi$$

$$=\frac{1}{T^2}\int_{-T}^{T}|f(x)|^2\,dx-\int_{-1}^{1}|f(x)|^2\,dx+\int_1^T\frac{2\,dx}{x^3}\int_{-x}^{x}|f(\xi)|^2\,d\xi$$

$$=O(1)+\int_1^T 2\,O(1)\frac{dx}{x^2}$$

$$=O(1). \tag{5.47}$$

Consequently

$$\left[\int_1^\infty+\int_{-\infty}^{-1}\right]\frac{|f(x)|^2}{x^2}\,dx \tag{5.48}$$

exists. It follows from this that

$$\psi_\mu(u)=\frac{1}{\pi}\,\underset{A\to\infty}{\mathrm{l.i.m.}}\int_{-A}^{A}f(x)\frac{\sin\mu x}{x}\,e^{iux}\,dx \tag{5.49}$$

exists, and that

$$\psi_\mu(u)=s(u+\mu)-s(u-\mu), \tag{5.50}$$

where

$$s(u)=\frac{1}{2\pi}\int_{-1}^{1}f(x)\frac{e^{iux}-1}{ix}\,dx+\frac{1}{2\pi}\,\underset{A\to\infty}{\mathrm{l.i.m.}}\left[\int_1^A+\int_{-A}^{-1}\right]\frac{f(x)\,e^{iux}}{ix}\,dx. \tag{5.51}$$

$s(u)$ has a somewhat artificial appearance, due to the fact that it is necessary to avoid the consequences of the vanishing of the denominator at the origin.

We shall see later, however, that we always actually work with $\psi_\mu(u)$ rather than with $s(u)$.

As a result of the Plancherel theory,

$$\frac{1}{2\mu}\int_{-\infty}^{\infty}|s(u+\mu)-s(u-\mu)|^2\,du=\frac{1}{\pi\mu}\int_{-\infty}^{\infty}|f(x)|^2\frac{\sin^2\mu x}{x^2}\,dx. \tag{5.52}$$

It follows from this by an immediate application of the Tauberian theorem associated with the names of Bochner, Hardy, Jacob, Littauer, and the author, and already proved in this section, that

$$\lim_{\mu\to 0}\frac{1}{2\mu}\int_{-\infty}^{\infty}|s(u+\mu)-s(u-\mu)|^2\,du=\lim_{T\to\infty}\frac{1}{2T}\int_{-T}^{T}|f(x)|^2\,dx. \tag{5.53}$$

The meaning of (5.53) is that *if $f(x)$ is quadratically summable over every finite range, and $f(x)/x$ is quadratically summable over any infinite range excluding the origin, then if either side of (5.53) exists, the other side exists and assumes the same value.*

This formula is worthy of some detailed attention. If $s(u)$ is of limited total variation, we shall always have

$$\frac{1}{2\mu}\int_{-\infty}^{\infty}|s(u+\mu)-s(u-\mu)|\,du\leq V(s). \tag{5.54}$$

Accordingly, if in addition

$$\lim_{T\to\infty}\frac{1}{2T}\int_{-T}^{T}|f(x)|^2\,dx\neq 0, \tag{5.55}$$

the function $s(u)$ cannot be uniformly continuous. Again, if

$$s(u)=A_n,\quad [\lambda_n<u<\lambda_{n+1}] \tag{5.56}$$

we shall have

$$\lim_{\mu\to 0}\frac{1}{2\mu}\int_{-\infty}^{\infty}|s(u+\mu)-s(u-\mu)|^2\,du=\Sigma|A_{n+1}-A_n|^2, \tag{5.57}$$

so that $\lim_{T\to\infty}\frac{1}{2T}\int_{-T}^{T}|f(x)|^2\,dx$ represents the sum of the squares of the moduli of the jumps of $s(u)$. Let it be noted that if $f(x)$ is a periodic function with the period 2π,

$$\lim_{T\to\infty}\frac{1}{2T}\int_{-T}^{T}|f(x)|^2\,dx=\frac{1}{2\pi}\int_{-\pi}^{\pi}|f(x)|^2\,dx, \tag{5.58}$$

while if

$$a_n=\frac{1}{2\pi}\int_{-\pi}^{\pi}f(x)\,e^{inx}\,dx, \tag{5.59}$$

then

$$\psi_\mu(u)=s(u+\mu)-s(u-\mu)=\sum_{[u-\mu]+1}^{[u+\mu]}a_n. \tag{5.60}$$

Thus our formula (5.53) is a generalization of the Parseval formula for the Fourier series, though it is not a direct generalization of the Parseval formula for the Fourier integral. For the Fourier integral,

$$\lim_{\mu\to 0}\frac{1}{2\mu}\int_{-\infty}^{\infty}|s(u+\mu)-s(u-\mu)|^2\,du=0, \tag{5.61}$$

although $s(u)$ exists, and indeed becomes the integral of the Fourier transform of $f(x)$. In this case, $s(u)$ is of limited total variation over every finite interval.

6. Bochner's generalization of harmonic analysis.

The study of the function $s(u)$ and its generalizations was first undertaken by Hahn, although, as we shall see later, on a basis insufficiently general to cover the needs of physics. The present author developed the theory for functions $f(x)$ with a finite mean square modulus, but the complete generalization of the theory is due to Bochner.

We have so far been interested in the problem of proceeding from $f(x)$ to $s(x)$. The question now arises, can we go backward, and determine $f(x)$ from $s(x)$? We should formally expect

20—29764. *Acta mathematica* 55. Imprimé le 8 avril 1930.

$$f(x) = \int_{-\infty}^{\infty} e^{-ixu}\, ds(u), \tag{6.01}$$

though the integral in question cannot be an ordinary Stieltjes integral, as $s(u)$ is not in general of limited total variation.

We may, however, develop this integral by a formal integration by parts, and we get

$$\int_{-A}^{A} e^{-ixu}\, ds(u) = e^{-iAx} s(A) - e^{iAx} s(-A) + ix \int_{-A}^{A} s(u)\, e^{-ixu}\, du$$

$$= e^{-iAx}\left[\frac{1}{2\pi}\int_{-1}^{1} f(\xi)\frac{e^{iA\xi}-1}{i\xi}\, d\xi + \underset{B\to\infty}{\text{l.i.m.}}\, \frac{1}{2\pi}\left[\int_{-B}^{-1} + \int_{1}^{B}\right] f(\xi)\frac{e^{iA\xi}}{i\xi}\, d\xi\right]$$

$$- e^{iAx}\left[\frac{1}{2\pi}\int_{-1}^{1} f(\xi)\frac{e^{-iA\xi}-1}{i\xi}\, d\xi + \underset{B\to\infty}{\text{l.i.m.}}\, \frac{1}{2\pi}\left[\int_{-B}^{-1} + \int_{1}^{B}\right] f(\xi)\frac{e^{-iA\xi}}{i\xi}\, d\xi\right]$$

$$+ ix \int_{-A}^{A} e^{-ixu}\, du \left[\frac{1}{2\pi}\int_{-1}^{1} f(\xi)\frac{e^{iu\xi}-1}{i\xi}\, d\xi + \underset{B\to\infty}{\text{l.i.m.}}\, \frac{1}{2\pi}\left[\int_{-B}^{-1} + \int_{1}^{B}\right] \frac{f(\xi)\, e^{iux}}{i\xi}\, d\xi\right]$$

$$= \underset{B\to\infty}{\text{l.i.m.}}\, \frac{1}{2\pi}\int_{-B}^{B} \frac{f(\xi)}{i\xi}\left[e^{iA(\xi-x)} - e^{-iA(\xi-x)} + ix\int_{-A}^{A} e^{iu(\xi-x)}\, du\right] d\xi$$

$$= \underset{B\to\infty}{\text{l.i.m.}}\, \frac{1}{\pi}\int_{-B}^{B} f(\xi)\frac{\sin A(\xi-x)}{\xi-x}\, d\xi. \tag{6.02}$$

Even this expression fails in general to converge in the mean as $A \to \infty$. A natural device to choose to compel the desired convergence in the mean is to replace this integral by its Cesaro sum, and to investigate the behavior of

$$\frac{1}{D}\int_{0}^{D} dA \int_{-A}^{A} e^{-ixu}\, ds(u) = \frac{1}{\pi D}\int_{-\infty}^{\infty} f(\xi)\frac{1-\cos D(\xi-x)}{(\xi-x)^2}\, d\xi. \tag{6.03}$$

This is the familiar Fejér expression for the partial Cesaro sum of a Fourier series, at least in form. The classical Fejér argument will prove that at any point of continuity of $f(x)$ we shall have

$$f(x) = \lim_{D\to\infty} \frac{1}{\pi D} \int_{-N}^{N} f(\xi) \frac{1-\cos D(\xi-x)}{(\xi-x)^2} d\xi, \tag{6.04}$$

and indeed, that this will in any case be true almost everywhere. To proceed to

$$\lim_{D\to\infty} \frac{1}{D} \int_{0}^{D} dA \int_{-A}^{A} e^{-ixu} \, ds(u) = f(x) \tag{6.05}$$

requires only the reflection that $\left[\int_{-\infty}^{-1} + \int_{1}^{\infty}\right] \frac{f(t)}{t^2} dt$ converges.

In a manner similar to that in which we have proved

$$\int_{-A}^{A} e^{-ixu} \, ds(u) = \underset{B\to\infty}{\text{l.i.m.}} \frac{1}{\pi} \int_{-B}^{B} f(\xi) \frac{\sin A(\xi-x)}{\xi-x} d\xi \tag{6.02}$$

we may show that

$$\int_{P}^{A} e^{-ixu} \, ds(u) = \underset{B\to\infty}{\text{l.i.m.}} \frac{1}{2\pi i} \int_{-B}^{B} f(\xi+x) \frac{e^{iA\xi} - e^{iP\xi}}{\xi} d\xi \tag{6.06}$$

as a function of A and P. Thus, except for an additive constant, $\int_{P}^{A} e^{-ixu} \, ds(u)$ bears to $f(x+\xi)$ the relation which $s(A)$ bears to $f(\xi)$. Similarly, to

$$\left.\begin{array}{ll} f(x) \pm f(t+x) \text{ there corresponds } \displaystyle\int_{P}^{A} (1 \pm e^{-itu}) \, ds(u) & \\ \text{and to} & \\ f(x) \pm if(t+x) \text{ there corresponds } \displaystyle\int_{P}^{A} (1 \pm ie^{-itu}) \, ds(u). & \end{array}\right\} \tag{6.07}$$

By an obvious linear combination of the formulae relating to these four functions separately (cf. (1.24)), we obtain

$$\lim_{\varepsilon\to 0}\frac{1}{2\varepsilon}\int_{-\infty}^{\infty}\left[\int_{u-\varepsilon}^{u+\varepsilon} e^{-itv}\,ds(v)\right][\bar{s}(u+\varepsilon)-\bar{s}(u-\varepsilon)]\,du$$

$$=\lim_{T\to\infty}\frac{1}{2T}\int_{-T}^{T} f(x+t)\bar{f}(x)\,dx=\varphi(t). \qquad (6.08)$$

Here, by definition,

$$\int_{u-\varepsilon}^{u+\varepsilon} e^{-itv}\,ds(v)=e^{-it(u+\varepsilon)}s(u+\varepsilon)-e^{-it(u-\varepsilon)}s(u-\varepsilon)+it\int_{u-\varepsilon}^{u+\varepsilon}s(v)\,e^{-itv}\,dv$$

$$=e^{-iut}(s(u+\varepsilon)-s(u-\varepsilon))$$

$$+\,it\left\{\int_{u}^{u+\varepsilon}[s(v)-s(u+\varepsilon)]\,e^{-itv}\,dv+\int_{u-\varepsilon}^{u}[s(v)-s(u-\varepsilon)]\,e^{itv}\,dv\right\}. \qquad (6.09)$$

However, we have

$$\frac{1}{\varepsilon}\left[\int_{u-\varepsilon}^{u+\varepsilon} e^{-itv}\,ds(v)-e^{-iut}(s(u+\varepsilon)-s(u-\varepsilon))\right]$$

$$=\underset{B\to\infty}{\text{l.i.m.}}\,\frac{1}{2\pi i\varepsilon}\int_{-B}^{B} f(\xi+t)\left[\frac{e^{i(u+\varepsilon)\xi}-e^{i(u-\varepsilon)\xi}}{\xi}-e^{-iut}\,\frac{e^{i(u+\varepsilon)(\xi+t)}-e^{i(u-\varepsilon)(\xi+t)}}{\xi+t}\right]d\xi$$

$$=\underset{B\to\infty}{\text{l.i.m.}}\,\frac{1}{2\pi i\varepsilon}\int_{-B}^{B} f(\xi+t)\left\{e^{i(u+\varepsilon)\xi}\left(\frac{1}{\xi}-\frac{e^{i\varepsilon t}}{\xi+t}\right)-e^{i(u-\varepsilon)\xi}\left(\frac{1}{\xi}-\frac{e^{-i\varepsilon t}}{\xi+t}\right)\right\}d\xi$$

$$=\underset{B\to\infty}{\text{l.i.m.}}\,\frac{1}{\pi}\int_{-B}^{B} f(\xi+t)\,e^{iu\xi}\left[\frac{\sin\varepsilon\xi}{\varepsilon\xi}-\frac{\sin\varepsilon(\xi+t)}{\varepsilon(\xi+t)}\right]d\xi. \qquad (6.10)$$

Now,

$$\left|\frac{\sin\varepsilon\xi}{\varepsilon\xi}-\frac{\sin\varepsilon(\xi+t)}{\varepsilon(\xi+t)}\right|$$

$$=\left|\frac{-2\xi\sin\frac{\varepsilon t}{2}\cos\varepsilon\left(\xi+\frac{t}{2}\right)+t\sin\varepsilon\xi}{\varepsilon\xi(\xi+t)}\right|\leq\left|\frac{2t}{\xi+t}\right|\text{ and also }\leq 2. \qquad (6.11)$$

Thus, by the use of a Fourier transformation, it follows at once that

$$\int_{-\infty}^{\infty} \frac{1}{\varepsilon^2} \left| \int_{u-\varepsilon}^{u+\varepsilon} e^{-itv}\, ds(v) - e^{-iut}(s(u+\varepsilon)-s(u-\varepsilon)) \right|^2 du = O(1). \tag{6.12}$$

Inasmuch as we may readily show that

$$\lim_{\varepsilon\to 0} \int_{-\infty}^{\infty} | s(u+\varepsilon)-s(u-\varepsilon) |^2\, du = 0, \tag{6.13}$$

and since

$$\begin{aligned} \varphi(t) &= \lim_{\varepsilon\to 0} \frac{1}{2\varepsilon} \int_{-\infty}^{\infty} e^{-itu} | s(u+\varepsilon)-s(u-\varepsilon) |^2\, du \\ &+ \lim_{\varepsilon\to 0} \frac{1}{2\varepsilon} \int_{-\infty}^{\infty} \left[\int_{u-\varepsilon}^{u+\varepsilon} e^{-itv}\, ds(v) - e^{-iut}(s(u+\varepsilon)-s(u-\varepsilon)) \right] [\bar{s}(u+\varepsilon)-\bar{s}(u-\varepsilon)]\, du, \end{aligned} \tag{6.14}$$

it follows by an elementary use of the Schwarz inequality that

$$\varphi(t) = \lim_{\varepsilon\to 0} \frac{1}{2\varepsilon} \int_{-\infty}^{\infty} e^{-itu} | s(u+\varepsilon)-s(u-\varepsilon) |^2\, du. \tag{6.15}$$

This formula holds in the same sense as (5.53) for each t independently.

We may deduce from (6.06) the analogous formula

$$\int_{-\lambda}^{\lambda} dy \int_{P}^{A} e^{-iyu}\, ds(u) = \int_{P}^{A} \frac{2\sin \lambda u}{u}\, ds(u) \tag{6.16}$$

by an easily justified inversion of the order of integration. Furthermore,

$$\begin{aligned} &\int_{-\lambda}^{\lambda} dy \mathop{\mathrm{l.i.m.}}_{B\to\infty} \frac{1}{2\pi i} \int_{-B}^{B} f(\xi+y) \frac{e^{iA\xi}-e^{iP\xi}}{\xi}\, d\xi \\ &\qquad = \mathop{\mathrm{l.i.m.}}_{B\to\infty} \frac{1}{2\pi i} \int_{-B}^{B} \frac{e^{iA\xi}-e^{iP\xi}}{\xi}\, d\xi \int_{-\lambda}^{\lambda} f(\xi+y)\, dy, \end{aligned} \tag{6.17}$$

as may be deduced from the fact that

$$\int_N^\infty \frac{|f(\xi+y)|^2}{\xi^2}\,d\xi \quad \text{and} \quad \int_{-\infty}^{-N} \frac{|f(\xi+y)|^2}{\xi^2}\,d\xi$$

tend uniformly to 0 with increasing N for all y in $(-\lambda, \lambda)$. Thus to

$$\int_{-\lambda}^{\lambda} f(x+y)\,dy \quad \text{there corresponds} \quad \int_P^A \frac{2\sin\lambda u}{u}\,ds(u)$$

in the same sense in which to $f(x)$ there corresponds $s(A)$. It follows from (5. 53) that

$$\lim_{\mu\to 0}\frac{1}{2\mu}\int_{-\infty}^{\infty}\left|\int_{A-\mu}^{A+\mu}\frac{2\sin\lambda u}{u}\,ds(u)\right|^2 dA = \lim_{T\to\infty}\frac{1}{2T}\int_{-T}^{T}\left|\int_{-\lambda}^{\lambda} f(x+y)\,dy\right|^2 dx. \tag{6. 18}$$

As in (6. 09),

$$\int_{A-\mu}^{A+\mu}\frac{2\sin\lambda u}{u}\,ds(u) = \frac{2\sin\lambda(A+\mu)}{A+\mu}\,s(A+\mu) - \frac{2\sin\lambda(A-\mu)}{A-\mu}\,s(A-\mu)$$

$$-\int_{A-\mu}^{A+\mu} s(u)\,\frac{d}{du}\left(\frac{2\sin\lambda u}{u}\right)du, \tag{6. 19}$$

and as in (6. 10),

$$\frac{1}{\mu}\int_{A-\mu}^{A+\mu}\frac{2\sin\lambda u}{u}\,ds(u) - \frac{2\sin\lambda A}{A\mu}\,[s(A+\mu)-s(A-\mu)]$$

$$=\frac{1}{\mu}\int_{A-\mu}^{A+\mu} ds(u)\int_{-\lambda}^{\lambda} e^{-i\sigma u}\,d\sigma - 2\int_{-\lambda}^{\lambda} e^{-i\sigma A}\,d\sigma\,[s(A+\mu)-s(A-\mu)]$$

$$=\int_{-\lambda}^{\lambda} d\sigma\,\underset{B\to\infty}{\text{l.i.m.}}\,\frac{1}{\pi}\int_{-B}^{B} f(\xi+\sigma)\,e^{iA\sigma}\left[\frac{\sin\mu\xi}{\mu\xi} - \frac{\sin\mu(\xi+\sigma)}{\mu(\xi+\sigma)}\right]d\xi \tag{6. 20}$$

because (6. 10) holds uniformly over a finite range of σ, because of the fact that

$$\left| \frac{\sin \mu\xi}{\mu\xi} - \frac{\sin \mu(\xi+\sigma)}{\mu(\xi-\sigma)} \right| = O(\xi^{-1}) \tag{6. 21}$$

uniformly over such a range. Hence, as in (6. 12), we may show that

$$\int_{-\infty}^{\infty} \frac{1}{\mu^2} \left| \int_{A-\mu}^{A+\mu} \frac{2 \sin \lambda u}{u} \, ds(u) - \frac{2 \sin \lambda A}{A} [s(A+\mu) - s(A-\mu)] \right|^2 dA = O(1); \tag{6. 22}$$

and by an argument exactly similar to that leading to (6. 15), we obtain

$$\lim_{T\to\infty} \frac{1}{2T} \int_{-T}^{T} \left| \int_{-\lambda}^{\lambda} f(x+y)\, dy \right|^2 dx = \lim_{\mu\to 0} \frac{1}{2\mu} \int_{-\infty}^{\infty} \frac{4 \sin^2 \lambda u}{u^2} \,| s(u+\mu) - s(u-\mu) |^2 \, du \tag{6. 23}$$

for each λ independently, in the same sense as that in which we have proved (5. 53) *and* (6. 15).

We now proceed to the part of the theory that is specifically Bochner's. We wish to discuss the harmonic analysis of functions which are no longer bounded on the average (for so we may interpret the finiteness of $\lim_{T\to\infty} \frac{1}{2T} \int_{-T}^{T} |f(x)|^2 dx$), but instead have on the average an algebraic rate of growth as the argument proceeds to $\pm\infty$. That is, we assume the existence of

$$\lim_{T\to\infty} \frac{1}{2T^n} \int_{-T}^{T} |f(x)|^2 \, dx \tag{6. 24}$$

and we shall take n to be a positive odd integer. By arguments following identically the lines laid down for $n=1$, we show that

$$\left[\int_{1}^{T} + \int_{-T}^{-1} \right] \frac{|f(x)|^2}{x^{n+1}} \, dx = O(1), \tag{6. 25}$$

and hence that

$$\left[\int_{1}^{\infty} + \int_{-\infty}^{-1} \right] \frac{|f(x)|^2}{x^{n+1}} \, dx \tag{6. 26}$$

exists. We can then show the existence of

$$s(u) = s_1(u) + s_2(u)$$

$$= \frac{1}{2\pi}\int_{-1}^{1} f(x)\left[\left(\int_0^u\right)^{\frac{n+1}{2}} e^{ivx}\,dv\right]dx + \frac{1}{2\pi}\,\underset{A\to\infty}{\text{l.i.m.}}\left[\int_1^A + \int_{-A}^{-1}\right] f(x)\,\frac{e^{iux}}{(ix)^{\frac{n+1}{2}}}\,dx. \qquad (6.\ 27)$$

If we now put

$$\Delta_\mu^{\frac{n+1}{2}}(s) = s\left(u+\mu\left(\frac{n+1}{2}\right)\right) - ns\left(u+\mu\left(\frac{n-3}{2}\right)\right) + \frac{n(n-1)}{2!}\,s\left(u+\mu\left(\frac{n-7}{2}\right)\right)$$

$$- \cdots \pm s\left(u-\mu\left(\frac{n+1}{2}\right)\right)$$

$$= \frac{2^{\frac{n+1}{2}}}{2\pi}\,\underset{A\to\infty}{\text{l.i.m.}}\int_{-A}^{A} f(x)\,e^{iux}\,\frac{\sin^{\frac{n+1}{2}}\mu x}{x^{\frac{n+1}{2}}}\,dx, \qquad (6.\ 28)$$

we can show by a Plancherel argument that

$$\frac{1}{\mu}\int_{-\infty}^{\infty}\left|\Delta_\mu^{\frac{n+1}{2}}(s(u))\right|^2 du = \frac{2^n}{\pi\mu}\int_{-\infty}^{\infty}|f(x)|^2\,\frac{\sin^{n+1}\mu x}{x^{n+1}}\,dx. \qquad (6.\ 29)$$

It is easy to show that for any function $F(x)$,

$$\frac{1}{T^n}\int_0^T F(x)\,dx = \frac{1}{T^n}\int_0^T \frac{F(x)}{x^{n-1}}\,x^{n-1}\,dx = \int_0^1 \frac{F(Ty)}{(Ty)^{n-1}}\,y^{n-1}\,dy. \qquad (6.\ 30)$$

Inasmuch as

$$\int_0^1 y^{n-1}\,dy = 1/n, \qquad (6.\ 31)$$

if we put

$$F(x)/nx^{n-1} = G(x), \qquad (6.\ 32)$$

the above integral is a mean of G, in the sense in which means enter into our general Tauberian theorem. Furthermore,

$$\frac{2^{n+1}}{\pi\mu}\int_0^\infty F(x)\,\frac{\sin^{n+1}\mu x}{x^{n+1}}\,dx = \frac{2^{n+1}n}{\pi\mu}\int_0^\infty G(x)\,\frac{\sin^{n+1}\mu x}{x^2}\,dx$$

$$= \frac{2^{n+1}n}{\pi}\int_0^\infty G(y/\mu)\,\frac{\sin^{n+1}y}{y^2}\,dy \qquad (6.\ 33)$$

and if we put

$$\frac{2^{n+1}n}{\pi}\int_0^\infty \frac{\sin^{n+1}y}{y^2}\,dy = P_n, \tag{6.34}$$

then

$$\frac{2^{n+1}}{\pi\mu P_n}\int_0^\infty F(x)\frac{\sin^{n+1}\mu x}{x^{n+1}}\,dx \tag{6.35}$$

is another mean of the quantity G. If we set

$$M_1(x)\,[\text{or } M_2(x)] = \begin{cases} n x^{n-1}; & [0 < x < 1] \\ 0 \quad ; & [x \geq 1] \end{cases} \quad M_2(x)\,[\text{or } M_1(x)] = \frac{2^{n+1}n}{\pi P_n}\frac{\sin^{n+1}x}{x^2}; \tag{6.36}$$

then

$$\int_0^\infty M_1(x)\,x^{iu}\,dx = \frac{n}{n+iu} \neq 0. \tag{6.37}$$

As to

$$\int_0^\infty M_2(x)\,x^{iu}\,dx,$$

we have

$$\int_0^\infty \sin^{n+1}x\; x^{iu-2}\,dx$$

$$= \int_0^\infty \frac{(n+1)!}{2^{n-1}}\left[\frac{\sin^2 x}{\frac{n-1}{2}!\,\frac{n+3}{2}!} - \frac{\sin^2 2x}{\frac{n-3}{2}!\,\frac{n+5}{2}!} + \frac{\sin^2 3x}{\frac{n-5}{2}!\,\frac{n+7}{2}!} - \cdots\right] x^{iu-2}\,dx.$$

Hence

$$\int_0^\infty M_2(x)\,x^{iu}\,dx = \frac{4n(n+1)!}{\pi P_n}\cdot$$

$$\left[\frac{1}{\frac{n-1}{2}!\,\frac{n+3}{2}!} - \frac{2^{iu-1}}{\frac{n-3}{2}!\,\frac{n+5}{2}!} + \frac{3^{iu-1}}{\frac{n-5}{2}!\,\frac{n+7}{2}!} - \cdots\right]\int_0^\infty \sin^2 x\; x^{iu-2}\,dx. \tag{6.38}$$

We have already seen that

$$\int_0^\infty \sin^2 x\, x^{iu-2}\, dx \neq 0. \qquad (6.39)$$

Thus the question of the possibility that

$$\int_0^\infty M_2(x)\, x^{iu}\, dx$$

should vanish depends on the possibility of the vanishing of

$$\frac{1}{\frac{n-1}{2}!\,\frac{n+3}{2}!} - \frac{2^{iu-1}}{\frac{n-3}{2}!\,\frac{n+5}{2}!} + \frac{3^{iu-1}}{\frac{n-5}{2}!\,\frac{n+7}{2}!} - \cdots. \qquad (6.40)$$

It is easy enough to prove that this cannot vanish for $n=1, 3, 5, 7, 9, 11$ but the author has not yet been able to produce a proof in the general case.

In case this expression does not vanish, we may apply our Tauberian theorem, replacing $f(x)$ by $|f(x)|^2+|f(-x)|^2$. We shall assume to begin with that $f(x)$ vanishes in the neighbourhood of the origin. Then

$$\lim_{T\to\infty} \frac{1}{2T^n} \int_{-T}^{T} |f(x)|^2\, dx = A \qquad (6.41)$$

and

$$\lim_{\mu\to 0} \frac{1}{P_n \mu} \int_{-\infty}^{\infty} \left| \Delta_\mu^{\frac{n+1}{2}} s(u) \right|^2 du = \lim_{\mu\to 0} \frac{2^n}{P_n \pi\mu} \int_{-\infty}^{\infty} |f(x)|^2 \frac{\sin^{n+1} \mu x}{x^{n+1}}\, dx \qquad (6.42)$$

are equivalent.

We have put $f(x)=0$ in the neighbourhood of the origin to be sure of the boundedness of

$$\frac{1}{2T^n} \int_{-T}^{T} |f(x)|^2\, dx \quad \text{and} \quad \frac{2^n}{P_n \pi\mu} \int_{-\infty}^{\infty} |f(x)|^2 \frac{\sin^{n+1} \mu x}{x^{n+1}}\, dx. \qquad (6.43)$$

In any case,

$$\lim_{T\to\infty} \frac{1}{2T^n} \int_{-T}^{T} |f(x)|^2\, dx$$

will not be changed if $f(x)$ is made to vanish in this neighbourhood. Moreover,

$$\frac{1}{\mu}\int_{-B}^{B}|f(x)|^2\frac{\sin^{n+1}\mu x}{x^{n+1}}\,dx \leq \mu^n\int_{-B}^{B}|f(x)|^2\,dx. \tag{6.44}$$

Thus

$$\lim_{\mu\to 0}\frac{1}{P_n\mu}\int_{-\infty}^{\infty}\left|\Delta_\mu^{\frac{n+1}{2}}s(u)\right|du$$

will not be changed either, and we may always write:

$$\lim_{\mu\to 0}\frac{1}{P_n\mu}\int_{-\infty}^{\infty}\left|\Delta_\mu^{\frac{n+1}{2}}s(u)\right|^2du = \lim_{T\to\infty}\frac{1}{2T^n}\int_{-T}^{T}|f(x)|^2\,dx. \tag{6.45}$$

We now come to the problem of returning from $s(u)$ to $f(x)$. We should expect formally

$$f(x) = \int_{-\infty}^{\infty}e^{-iux}\,d^{\frac{n+1}{2}}s(u)/du^{\frac{n-1}{2}} \tag{6.46}$$

Again, we need to interpret

$$\int_{-\infty}^{\infty}e^{-iux}\,d^{\frac{n+1}{2}}s(u)/du^{\frac{n-1}{2}}$$

by an integration by parts, and again some form of summation is necessary to get a more manageable expression. One method is the following: Let us replace

$$\int_{-\infty}^{\infty}e^{-iux}\,d^{\frac{n+1}{2}}s(u)/du^{\frac{n-1}{2}}$$

by

$$\int_{-\infty}^{\infty}e^{-\lambda u^2-iux}\,d^{\frac{n+1}{2}}s(u)/du^{\frac{n-1}{2}} \tag{6.47}$$

and let us investigate its behavior as $\lambda \to 0$. This expression is to be interpreted by $\frac{n+1}{2}$ formal integrations by parts, which convert it to

$$(-1)^{\frac{n+1}{2}} \int_{-\infty}^{\infty} s_2(u) d^{\frac{n+1}{2}} (e^{-\lambda u^2 - iux}) / du^{\frac{n+1}{2}} du. \tag{6.48}$$

It then becomes obvious that this expression will have the same limit in the mean as

$$\int_{-B}^{B} s_2(u) (ix)^{\frac{n+1}{2}} e^{-iux} du \qquad [B \to \infty] \tag{6.49}$$

It here will, however, be $f(x)$ if $|x| > A$, and will be 0 otherwise. To see this, we need only compare $\dfrac{d^{\frac{n+1}{2}}}{dx^{\frac{n+1}{2}}} (e^{-\lambda u^2 - iux})$ with $\dfrac{d^{\frac{n+1}{2}}}{dx^{\frac{n+1}{2}}} e^{-iux}$. A similar result holds for $s_1(u)$, and the final result is that over any finite interval not including the origin,

$$f(x) = \underset{\lambda \to 0}{\text{l.i.m.}} \int_{-\infty}^{\infty} e^{-\lambda u^2 - iux} d^{\frac{n+1}{2}} s(u) / du^{\frac{n-1}{2}} \tag{6.50}$$

7. The Hahn generalization of harmonic analysis.

Up to the present we have concerned ourselves rather with questions of convergence in the mean than of ordinary convergence. Retaining our previous notation, in the case where $n=1$, we may raise the further questions: when does

$$\int_{-A}^{A} e^{-iux} ds(u)$$

exist for all u as an ordinary Stieltjes integral, rather than a generalized Stieltjes integral such as we have treated in the last section? When does

$$\lim_{A \to \infty} \int_{-A}^{A} e^{-iux} ds(u)$$

converge in the ordinary sense? These questions furnish the vital link between the generalized harmonic analysis of Hahn, and that developed here. In the second problem of this section, it is at present not likely that we can obtain conditions that are both necessary and sufficient. The necessary and sufficient answer to the first question is manifestly that $s(u)$ should be of limited total variation over any finite interval. We have

$$s(v)-s(u) = \underset{B\to\infty}{\text{l.i.m.}} \frac{1}{2\pi}\int_{-B}^{B} f(x)\frac{e^{ivx}-e^{iux}}{ix}\,dx$$

$$= \underset{B\to\infty}{\text{l.i.m.}} \frac{1}{2\pi}\int_{-B}^{B} \frac{e^{ivx}-e^{iux}}{ix}\,d\int_{0}^{x} f(\xi)\,d\xi$$

$$= \underset{B\to\infty}{\text{l.i.m.}} \frac{1}{2\pi}\left\{\frac{e^{ivB}-e^{iuB}}{iB}\int_{0}^{B} f(\xi)\,d\xi + \frac{e^{-ivB}-e^{-iuB}}{iB}\int_{0}^{-B} f(\xi)\,d\xi\right.$$

$$\left. + i\int_{-B}^{B}\left[\frac{ivx-1}{x^2}e^{ivx} - \frac{iux-1}{x^2}e^{iux}\right]dx\int_{0}^{x} f(\xi)\,d\xi\right\}. \tag{7.01}$$

If now we assume that

$$F(x) = \int_{0}^{x} f(\xi)\,d\xi = O(x^{-1}), \tag{7.02}$$

it follows that

$$s(v)-s(u) = i\int_{-\infty}^{\infty} F(x)\left[\frac{ivx-1}{x^2}e^{ivx} - \frac{iux-1}{x^2}e^{iux}\right]dx. \tag{7.03}$$

Inasmuch as by our assumptions

$$\int_{1}^{\infty}\frac{F(x)}{x^2}e^{iux}\,dx \quad\text{and}\quad \int_{1}^{\infty}\frac{F(x)}{x}e^{iux}\,dx$$

converge and represent functions with quadratically summable derivatives, $s(u)$ is of limited total variation over any finite region. Thus (7.02) is a sufficient condition for the answer to our first question to be in the affirmative.

Let us now suppose our first question affirmatively answered. The formal integration by parts by which we have defined $\int_{-A}^{A} e^{-iux}\,ds(u)$ in the general case may now be carried out, and all our quasi-Stieltjes integrals become ordinary Stieltjes integrals. If, on the other hand, $s(u)$ is not of limited total variation, Hahn adopts a generalized definition of the Stieltjes integral identical in content with that here given. In either case, we have already seen that

$$\int_{-A}^{A} e^{-ixu}\, ds(u) = \underset{B\to\infty}{\text{l.i.m.}}\, \frac{1}{\pi}\int_{-B}^{B} f(\xi)\frac{\sin A(\xi-x)}{\xi-x}\, d\xi, \tag{7.04}$$

and that

$$f(x) = \lim_{A\to\infty} \frac{1}{2\pi A}\int_{-\infty}^{\infty} f(\xi)\frac{1-\cos 2A(\xi-x)}{(\xi-x)^2}\, d\xi. \tag{7.05}$$

Moreover,

$$\underset{B\to\infty}{\text{l.i.m.}}\, \frac{1}{2\pi A}\int_{-B}^{B} f(\xi)\frac{1-\cos 2A(\xi-x)-2A(\xi-x)\sin A(\xi-x)}{(\xi-x)^2}\, d\xi$$

$$= \frac{1}{2\pi}\int_{-\infty}^{\infty} f\left(\xi+\frac{w}{A}\right)\frac{1-\cos 2w-2w\sin w}{w^2}\, dw$$

$$= \frac{1}{2\pi}\int_{-\infty}^{\infty}\left[f\left(\xi+\frac{w}{A}\right)-f(\xi)\right]\frac{1-\cos 2w-2w\sin w}{w^2}\, dw$$

$$+\frac{f(\xi)}{2\pi}\int_{-\infty}^{\infty}\frac{1-\cos 2w-2w\sin w}{w^2}\, dw$$

$$= \frac{1}{2\pi}\int_{-D}^{D}\left[f\left(\xi+\frac{w}{A}\right)-f(\xi)\right]\frac{1-\cos 2w-2w\sin w}{w^2}\, dw$$

$$+\frac{1}{2\pi}\left[\int_{-\infty}^{-D}+\int_{D}^{\infty}\right]\left[f\left(\xi+\frac{w}{A}\right)-f(\xi)\right]\frac{1-\cos 2w-2w\sin w}{w^2}\, dw. \tag{7.06}$$

Thus a sufficient condition for

$$\int_{-\infty}^{\infty} e^{-ixu}\, ds(u) = f(x) \tag{7.07}$$

is that $f(x)$ should satisfy locally one of the sufficient conditions for the convergence of a Fourier series, and that

$$\left[\int_{-\infty}^{-1} + \int_{1}^{\infty}\right]\left|\frac{f(w)}{w}\right| dw \text{ converge}$$

or what is the same, that

$$\int_{-\infty}^{\infty}\left|\frac{f(w)}{\sqrt{1+w^2}}\right| dw \tag{7.08}$$

exist. This condition thus constitutes a sufficient solution of the Hahn problem. To such a function we may add any function with a convergent Fourier series, without destroying the fact that it solves Hahn's problem.

Another condition under which (7.07) holds is that $f(x)$ should be of the form

$$g(x) \cdot h(x)$$

where $g(x)$ is periodic, and $h(x)$ bounded and monotone at $\pm\infty$, and that $f(x)$ should satisfy one of the sufficient conditions for the convergence of a Fourier series to its function. As in the previous case, the proof of his assertion made by Hahn depends simply on the fact that the second term of the last line of (7.06) will then vanish with increasing D, while the first term is asymptotically equal to the difference between the Cesaro and ordinary partial sums of the Fourier integral of the function $f(x)$ mutilated by being made to vanish outside the interval $(-D, D)$. It then follows from the Fejér theorem and the fact that the conditions for the convergence of such a Fourier integral and of a Fourier series are the same, that Hahn's theorem holds.

The class of functions for which $s(u)$ exists as an ordinary Stieltjes integral is too narrow to cover the physically interesting cases of continuous spectra. To see this, let $f(x)$ be a function for which $\varphi(x)$ and $S(u)$ exist, and let

$$f^{(\eta)}(x) = \frac{1}{2\eta}\cdot\int_{-\eta}^{\eta} f(x+\xi)\, d\xi. \tag{7.09}$$

Let us suppose that

$$\varphi^{(\eta)}(x) = \lim_{T\to\infty} \frac{1}{2T} \int_{-T}^{T} f^{(\eta)}(x+t)\bar{f}^{(\eta)}(t)\,dt \tag{7. 10}$$

exists for every x, and hence that

$$S^{(\eta)}(u) = \frac{1}{2\pi} \int_{-\infty}^{\infty} \varphi^{(\eta)}(x) \frac{e^{iux}-1}{ix}\,dx \tag{7. 11}$$

exists. Then

$$\begin{aligned}
\varphi^{(\eta)}(x) &= \lim_{T\to\infty} \frac{1}{2T}\frac{1}{4\eta^2} \int_{-\eta}^{\eta} d\xi \int_{-\eta}^{\eta} d\zeta \int_{-T}^{T} f(x+t+\xi)\bar{f}(t+\zeta)\,dt \\
&= \frac{1}{4\eta^2} \int_{-\eta}^{\eta} d\xi \int_{-\eta}^{\eta} d\zeta\, \varphi(x+\xi-\zeta) \\
&= \frac{1}{4\eta^2} \int_{-\eta}^{\eta} d\xi \int_{-\eta-\xi}^{\eta-\xi} \varphi(x-y)\,dy \\
&= \frac{1}{4\eta^2} \left\{ \eta \int_{-2\eta}^{0} \varphi(x-y)\,dy + \eta \int_{0}^{2\eta} \varphi(x-y)\,dy - \int_{-\eta}^{\eta} \xi[\varphi(x+\eta+\xi)-\varphi(x-\eta+\xi)]\,d\xi \right\} \\
&= \frac{1}{4\eta^2} \int_{-2\eta}^{2\eta} \varphi(x-y)(2\eta-|y|)\,dy;
\end{aligned} \tag{7. 12}$$

$$\begin{aligned}
S^{(\eta)}(u+\mu) - S^{(\eta)}(u-\mu) &= \frac{1}{\pi} \operatorname*{l.i.m.}_{A\to\infty} \int_{-A}^{A} \varphi^{(\eta)}(x) \frac{\sin \mu x}{x} e^{iux}\,dx \\
&= \operatorname*{l.i.m.}_{A\to\infty} \frac{1}{4\pi\eta^2} \int_{-2\eta}^{2\eta} (2\eta-|y|)\,dy \int_{-A-y}^{A-y} \varphi(x-y) \frac{\sin \mu x}{x} e^{iux}\,dx \\
&= \operatorname*{l.i.m.}_{A\to\infty} \frac{1}{4\pi\eta^2} \int_{-A}^{A} \varphi(x)dx \int_{-2\eta}^{2\eta} (2\eta-|y|) \frac{\sin \mu(x+y)}{x+y} e^{iu(x+y)}\,dy
\end{aligned}$$

$$= \underset{A\to\infty}{\mathrm{l.i.m.}} \frac{1}{8\pi\eta^2}\int_{-A}^{A} \varphi(x)\,dx \int_{u-\mu}^{u+\mu} dw \int_{-2\eta}^{2\eta} (2\eta - |y|)\, e^{iw(x+y)}\,dy$$

$$= \underset{A\to\infty}{\mathrm{l.i.m.}} \frac{1}{4\pi\eta^2}\int_{-A}^{A} \varphi(x)\,dx \int_{u-\mu}^{u+\mu} \frac{e^{iwx}(1-\cos 2w\eta)}{w^2}\,dw$$

$$= \underset{A\to\infty}{\mathrm{l.i.m.}} \frac{1}{4\pi\eta^2}\int_{u-\mu}^{u+\mu} \frac{1-\cos 2w\eta}{w^2}\,dw \int_{-A}^{A} \varphi(x)\frac{e^{iwx}-1}{ix}\,dx$$

$$= \frac{1}{4\pi\eta^2}\int_{u-\mu}^{u+\mu} \frac{1-\cos 2w\eta}{w^2}\,dw \int_{-\infty}^{\infty} \varphi(x)\frac{e^{iwx}-1}{ix}\,dx$$

$$= \frac{1}{2\eta^2}\int_{u-\mu}^{u+\mu} \frac{1-\cos 2w\eta}{w^2}\,dS(w) = \int_{u-\mu}^{u+\mu} \frac{\sin^2 \eta w}{\eta^2 w^2}\,dS(w); \qquad (7.13)$$

$$s^{(\eta)}(u+\varepsilon) - s^{(\eta)}(u-\varepsilon) = \underset{A\to\infty}{\mathrm{l.i.m.}} \frac{1}{\pi}\int_{-A}^{A} f^{(\eta)}(x)\frac{\sin \varepsilon x}{x} e^{iux}\,dx$$

$$= \underset{A\to\infty}{\mathrm{l.i.m.}} \frac{1}{2\pi\eta}\int_{-\eta}^{\eta} d\xi \int_{-A}^{A} f(x+\xi)\frac{\sin \varepsilon x}{x} e^{iux}\,dx$$

$$= \frac{1}{2\eta}\int_{-\eta}^{\eta} d\xi \int_{u-\varepsilon}^{u+\varepsilon} e^{-i\xi u}\,ds(u) \ \text{[by (6.06)]}. \qquad (7.14)$$

Although this is not in general an ordinary Stieltjes integral, we may integrate by parts and then invert the order of integration, and thus obtain

$$s^{(\eta)}(u+\varepsilon) - s^{(\eta)}(u-\varepsilon) = \frac{1}{2\eta}\int_{u-\varepsilon}^{u+\varepsilon} ds(u) \int_{-\eta}^{\eta} e^{-i\xi u}\,d\xi = -\int_{u-\varepsilon}^{u+\varepsilon} \frac{\sin \eta u}{\eta u}\,ds(u). \qquad (7.15)$$

Hence, by (5.53) and (6.23),

$$\varphi^{(\eta)}(0) = \lim_{T\to\infty}\frac{1}{2T}\int_{-T}^{T} |f^{(\eta)}(x)|^2\,dx = \lim_{\varepsilon\to 0}\frac{1}{2\varepsilon}\int_{-\infty}^{\infty} \frac{\sin^2 \eta u}{\eta^2 u^2}\,|s(u+\varepsilon) - s(u-\varepsilon)|^2\,du. \qquad (7.16)$$

By (7. 14),

$$\varphi^{(\eta)}(0) = \frac{1}{4\eta^2}\int_{-2\eta}^{2\eta} \varphi(\xi)(2\eta - |\xi|)\, d\xi$$

$$= \frac{1}{4\eta^2}\int_{0}^{2\eta} 2\lambda\, d\lambda \left[\frac{1}{2\lambda}\int_{-\lambda}^{\lambda} \varphi(\xi)\, d\xi\right], \tag{7. 17}$$

which in combination with (5. 47) leads us to

$$\lim_{\eta\to 0} \varphi^{(\eta)}(0) = S(\infty) - S(-\infty). \tag{7. 18}$$

In other words,

$$\lim_{\eta\to 0}\lim_{\varepsilon\to 0} \frac{1}{2\varepsilon}\int_{-\infty}^{\infty} \frac{\sin^2 \eta u}{\eta^2 u^2} |s(u+\varepsilon) - s(u-\varepsilon)|^2\, du = S(\infty) - S(-\infty). \tag{7. 19}$$

If we now assume that

$$\lim_{\eta\to 0} \varphi^{(\eta)}(0) = \varphi(0), \tag{7. 20}$$

it follows from (7. 21), (7. 22), and (5. 54) that

$$\lim_{\eta\to 0}\lim_{\varepsilon\to 0} \frac{1}{2\varepsilon}\int_{-\infty}^{\infty} \left[1 - \frac{\sin^2 \eta u}{\eta^2 u^2}\right] |s(u+\varepsilon) - s(u-\varepsilon)|^2\, du = 0, \tag{7. 21}$$

or since $\sin \eta u/\eta u$ tends to 1 as $\eta \to 0$

$$\lim_{A\to\infty}\lim_{\varepsilon\to 0} \frac{1}{2\varepsilon}\left[\int_{A}^{\infty} + \int_{-\infty}^{-A}\right] |s(u+\varepsilon) - s(u-\varepsilon)|^2\, du = 0. \tag{7. 22}$$

If $\varphi(x)$ is a continuous function, not only shall we have (7. 20) as a consequence of (7. 17), but all the functions $\varphi^{(\eta)}(x)$ will be continuous, as follows from (7. 12). Then, by (6. 15), we shall have

$$\lim_{\varepsilon\to 0} \frac{1}{2\varepsilon}\int_{-\infty}^{\infty} e^{-iux} |s(u+\varepsilon) - s(u-\varepsilon)|^2\, du = \int_{-\infty}^{\infty} e^{-iux}\, d\, S(u). \tag{7. 23}$$

It follows at once that if $P(u)$ is a trigonometrical polynomial, or the uniform limit of such a polynomial,

$$\lim_{\varepsilon\to 0}\frac{1}{2\varepsilon}\int_{-\infty}^{\infty} P(u)\,|s(u+\varepsilon)-s(u-\varepsilon)|^2\,du=\int_{-\infty}^{\infty} P(u)\,dS(u). \tag{7.24}$$

For the transition to the case where $P(u)$ is a uniform limit of a polynomial we need only make appeal to the boundedness of

$$\lim_{\varepsilon\to 0}\frac{1}{2\varepsilon}\int_{-\infty}^{\infty} |s(u+\varepsilon)-s(u-\varepsilon)|^2\,du. \tag{7.25}$$

Hence, by Fejér's theorem, $P(u)$ may be any continuous periodic function, and because of (7.22), any continuous function differing from 0 only over a finite range, as the change in the left side of this expression due to making $P(u)$ artificially periodic tends to 0 as the period increases.

It follows at once that under the hypotheses:

(a) $\varphi^{(\eta)}(x)$ exists for every x and η;

(b) $\varphi(x)$ is continuous;

if

$$\lim_{\varepsilon\to 0}\frac{1}{2\varepsilon}\int_{\alpha}^{\beta} |s(u+\varepsilon)-s(u-\varepsilon)|^2\,du=0, \tag{7.26}$$

over any interval, $S(u)$ is a constant over any interior interval. Thus if $s(u)$ is of limited total variation over any finite interval, and is continuous, $S(u)$ reduces to a constant, and the spectrum of $f(x)$ vanishes. In other words, the very natural hypotheses (a) and (b) are inconsistent with the existence of a continuous spectrum, provided $s(u)$ is of limited total variation. To see this, we need only notice that almost everywhere

$$S(u_2)-S(u_1)=\lim_{\varepsilon\to 0}\frac{1}{2\varepsilon}\int_{u_1}^{u_2} |s(u+\varepsilon)-s(u-\varepsilon)|^2\,du$$

$$=\text{sum of squares of jumps of } S(u) \text{ between } u_1 \text{ and } u_2. \tag{7.27}$$

Thus the expansion of $f(x)$ in an ordinary Stieltjes integral is not adequate to the discussion of such continuous spectra as occur in physics, inasmuch, as in these cases, as we shall see in section 13, conditions analogous to (a) and (b) are fulfilled.

CHAPTER III.

8. Harmonic analysis in more than one dimension.

The elementary function of harmonic analysis in one dimension is e^{iux}. In n dimensions, this is replaced by

$$e^{i(u_1x_1+\cdots+u_nx_n)}$$

which we may write vectorially

$$e^{i(U\cdot X)},$$

where the vector X represents the argument of the function to be analyzed, and the vector U the vectorial frequency. If we keep the term $(U . X)$ invariant, and X varies cogrediently, U varies contragrediently. Thus the familiar duality relation between Fourier transforms is intimately connected with the point-plane duality of geometry. This is why the relation between position-coordinates and momentum-coordinates in modern quantum physics appears as a Fourier duality, while the same relation appears in the theory of relativity as the relation between a certain cogredient tensor and a certain contragredient tensor.

Practically the whole generalized theory of harmonic analysis so far developed is susceptible to a generalization to n dimensions. This generalization has been carried out by Mr. A. C. Berry, who has been kind enough to furnish me with the following summary of his Harvard doctoral thesis.

It is necessary to introduce certain notations at once and to make a few preliminary remarks. Let there be given a real n-dimensional space and, in it, some fixed reference point, or origin, O. If X is an arbitrary point of this space, the symbol X shall be used to denote not only this point, but also any real n-dimensional vector equivalent to the directed line segment OX. Let $f(X)$ be any complex, measurable function defined for all such real arguments X. Hereafter it will be assumed that all functions with which we start satisfy these requirements. If R is any measurable point-set, then

$$\int_R f(X)\, d\, V_X \tag{8.01}$$

shall mean the n-dimensional volume integral, in the sense of Lebesgue, of $f(X)$ taken over the region R. Since, in general, n-dimensional »spheres» will be employed as regions of integration, it will be convenient to use the notation $(r;\, X)$ to signify a sphere of radius r and having its center at X. The vector interpretation of X enables us to write

$$\int_{(r;\,Y)} f(X)\, d\, V_X = \int_{(r;\,O)} f(X+Y)\, d\, V_X. \tag{8.02}$$

The »volume» or measure, of an n-dimensional sphere of radius r is known to be

$$\frac{\pi^{\frac{n}{2}}\, r^n}{\Gamma\left(\frac{n}{2}+1\right)} \tag{8.03}$$

which quantity, hereafter, shall be denoted by the symbol $v(r)$. Thus the average of a function $f(x)$ over a sphere of radius r about the point Y is

$$\frac{1}{v(r)} \int_{(r;\,Y)} f(X)\, d\, V_X. \tag{8.04}$$

Corresponding to the theorem, in one dimension, that a function is almost everywhere the derivative of its integral, there is here the fact that, for almost every Y,

$$\lim_{r\to 0} \frac{1}{v(r)} \int_{(r;\,Y)} f(X)\, d\, V_X = \lim_{r\to 0} \frac{1}{v(r)} \int_{(r;\,O)} f(X+Y)\, d\, V_X = f(Y). \tag{8.05}$$

For any positive integer m it readily follows that, almost everywhere,

$$\begin{aligned} &\lim_{r\to 0} \frac{1}{[v(r)]^m} \int_{(r;\,Y)} d\, V_{X_m} \cdots \int_{(r;\,X_3)} d\, V_{X_2} \int_{(r;\,X_2)} f(X_1)\, d\, V_{X_1} \\ &= \lim_{r\to 0} \frac{1}{[v(r)]^m} \int_{(r;\,O)} d\, V_{X_m} \cdots \int_{(r;\,O)} d\, V_{X_2} \int_{(r;\,O)} f(X_1+X_2+\cdots+X_m+Y)\, d\, V_{X_1} \\ &= f(Y). \end{aligned} \tag{8.06}$$

The classical Stieltjes integral, for functions of a single variable,

$$\int_{-\infty}^{\infty} f(x)\, d\varphi(x),$$

may be defined under suitable conditions as the following limit:

$$\lim_{r\to 0} \frac{1}{2r} \int_{-\infty}^{\infty} f(x)\{\varphi(x+r)-\varphi(x-r)\}\, dx. \tag{8.07}$$

If one denotes by $(r;\, x)$ that interval of length $2r$ which has its center at the point x and if one constructs the following function of an interval:

$$M(r;\, x) = \varphi(x+r)-\varphi(x-r), \tag{8.08}$$

then this limit may be written in the form

$$\lim_{r\to 0} \frac{1}{2r} \int_{-\infty}^{\infty} f(x)\, M(r;\, x)\, dx, \tag{8.09}$$

which may be called the Stieltjes integral of the point-function f with respect to the point-set-function M. Proceding by analogy we shall call the expression

$$\lim_{r\to 0} \frac{1}{v(r)} \int_{\infty} f(X)\, M(r;\, X)\, d\, V_X, \tag{8.10}$$

when it exists, the Stieltjes integral of the n-dimensional point-function f with respect to the region-function M, and shall denote it by the symbol

$$\int_{\infty} f(X)\, d_X M. \tag{8.11}$$

It will be necessary to introduce a generalization of this integral in order to handle certain region-functions M to appear later. If the expression

$$\lim_{r\to 0} \frac{1}{[v(r)]^m} \int_{\infty} f(X)\, M(r;\, X)\, d\, V_X \tag{8.12}$$

exists it shall be called the mth Stieltjes integral of f with respect to M and will be denoted by

$$\int_{\infty} f(X)\, d_X^m M. \tag{8.13}$$

A function $f(X)$ shall be said to be quadratically summable over all space, or q. s. over ∞, provided it satisfies the requirements laid down above and provided that

$$\int_{\infty} |f(X)|^2\, d\, V_X \tag{8.14}$$

exists. A one-parameter family of functions $f_n(X)$, each of which is q. s. over ∞, is said to converge in mean to a function $f(X)$, also q. s. over ∞, as $n \to \infty$, if

$$\lim_{\lambda \to \infty} \int_{\infty} |f(X) - f_n(X)|^2\, d\, V_X = 0. \tag{8.15}$$

We are now in a position to discuss the harmonic analysis of a given function $f(X)$. The fundamental theorem is, of course, Plancherel's theorem on the Fourier transform. In n dimensions this reads as follows:

If $f(X)$ is q. s. over ∞, there exists its Fourier transform

$$g(U) = \underset{R \to \infty}{\text{l.i.m.}}\, \frac{1}{(2\pi)^{n/2}} \int_R f(X)\, e^{i(X \cdot U)}\, d\, V_X, \tag{8.16}$$

where the expanding region R is selected from an arbitrary one-parameter family of regions which are such that ultimately R covers and continues to cover almost any given point of space. All such transforms are equivalent; i.e. any pair of such functions can differ at most on a set of zero measure. Furthermore, $g(U)$ is q. s. over ∞ and satisfies the equality

$$\int_{\infty} |g(U)|^2\, d\, V_U = \int_{\infty} |f(X)|^2\, d\, V_X. \tag{8.17}$$

For any integer $m \geq 1$, there exists as an absolutely convergent integral

$$\int\limits_{(r;\,0)} d\,V_{T_m} \cdots \int\limits_{(r;\,0)} d\,V_{T_2} \int\limits_{(r;\,0)} g(T_1+T_2+\cdots+T_m+U)\,d\,V_{T_1}$$

$$= \frac{1}{(2\,\pi)^{n/2}} \int\limits_{\infty} f(X) \left\{ \int\limits_{(r;\,0)} e^{i(X\cdot T)}\,d\,V_T \right\}^m e^{i(X\cdot U)}\,d\,V_X, \tag{8. 18}$$

which yields the following explicit formula for $g(U)$:

$$g(U) = \lim_{r\to 0} \frac{1}{(2\,\pi)^{n/2}} \cdot \frac{1}{[v(r)]^m} \int\limits_{\infty} f(X) \left\{ \int\limits_{(r;\,0)} e^{i(X\cdot T)}\,d\,V_T \right\}^m e^{i(X\cdot U)}\,d\,V_X;\quad m=1,2,\ldots; \tag{8. 19}$$

this limit existing for almost every U. Conversely, $g(U)$ possesses a Fourier transform and this is $f(-X)$:

$$f(X) = \underset{R\to\infty}{\text{l.i.m.}}\, \frac{1}{(2\,\pi)^{n/2}} \int\limits_{R} g(U)\,e^{-i(X\cdot U)}\,d\,V_U. \tag{8. 20}$$

An equivalent statement is that $\bar{f}(X)$ is the Fourier transform of $\bar{g}(U)$. As above, we may write explicitly, for almost every X,

$$f(X) = \lim_{r\to 0} \frac{1}{(2\,\pi)^{n/2}} \cdot \frac{1}{[v(r)]^m} \int\limits_{\infty} g(U) \left\{ \int\limits_{(r;\,0)} e^{-i(U\cdot T)}\,d\,V_T \right\}^m e^{-i(X\cdot U)}\,d\,V_U;\quad m=1,2,\ldots. \tag{8. 21}$$

Finally, if $f_1(X)$ and $f_2(X)$ are each q. s. over ∞, and if $g_1(U)$ and $g_2(U)$ are their respective Fourier transforms, then

$$\int\limits_{\infty} f_1(X)\,g_2(X)\,d\,V_X = \int\limits_{\infty} f_2(U)\,g_1(U)\,d\,V_U. \tag{8. 22}$$

We are thus possessed of a harmonic analysis of any quadratically summable function. For a given such function $f(X)$ this consists in associating with almost every vector frequency U a complex amplitude $g(U)$. It is suggestive to imagine this $g(U)$ as the density of a complex mass distribution in the U space. The converse procedure by which we rebuild the given $f(X)$ from this mass distribution may be described formally as follows: We begin by multiplying the density of this distribution by the factor,

$$e^{-i(X\cdot U)}$$

thus altering the complex phase of the density in an simply periodic fashion. This done, we calculate the total mass of the resulting distribution and find it to be $f(X)$.

Now, there exists a very important class of functions to which exactly this harmonic analysis cannot be given, namely the n-tuply periodic functions. If such a function $f(X)$ be also q. s. over an arbitrary finite region, it is known to possess an n-tuple Fourier series representation:

$$f(X) \sim \Sigma a_k e^{i(X \cdot U_k)}; \tag{8. 23}$$

where the summation is effected for all points U_k which are vertices of a certain rectangular network. This analysis can also be interpreted as a mass distribution. Here, however, the spread is not continuous but consists of masses a_k concentrated at the corresponding frequency points U_k. Yet the process by which $f(X)$ is reobtained is again that of calculating the total mass of a distribution.

If one seeks a uniform method of treating these two types of mass spreads, one naturally is led to construct a region-function: the total mass in an arbitrary region. Knowledge of this function is equivalent to knowledge of the particular distribution in question. The advantage derived in employing it is that it is of the same order of magnitude for the various types of spreads whereas the densities are not. We shall see that this region-function can readily be calculated. To effect a return to the given function $f(X)$ we shall employ the Stieltjes integral which will simultaneously correctly modify the complex phases of the spread and determine the total resulting mass. Let us note how easily all this is carried out.

However, we can handle at once a more general problem. Let $f(X)$ be a function such that, for some integer m, the product

$$f(X)\left\{\int\limits_{(r;\,O)} e^{i(X \cdot T)}\, d\, V_T\right\}^m \tag{8. 24}$$

is q. s. over ∞. Let the Fourier transform of this product be denoted by

$$M^{(m)}(r;\; U) = \underset{R \to \infty}{\text{l.i.m.}} \frac{1}{(2\,\pi)^{n/2}} \int\limits_{r} f(X) \left\{\int\limits_{(r;\,O)} e^{i(X \cdot T)}\, d\, V_T\right\}^m e^{i(X \cdot U)}\, d\, V_X. \tag{8. 25}$$

Hence

23 — 29764. *Acta mathematica*. 55. Imprimé le 14 avril 1930.

$$f(X)\left\{\int_{(r;\,O)} e^{i(X\cdot T)}\,d\,V_T\right\}^m = \underset{R\to\infty}{\text{l.i.m.}}\ \frac{1}{(2\,\pi)^{n/2}}\int_R e^{-i(X\cdot U)}\,M^{(m)}(r;\,U)\,d\,V_U. \tag{8. 26}$$

Integrating the above expression we obtain the result:

$$\int_{(s;\,Y)} f(X)\left\{\int_{(r;\,O)} e^{i(X\cdot T)}\,d\,V_T\right\}^m d\,V_X$$

$$= \frac{1}{(2\,\pi)^{n/2}}\int_\infty\left\{\int_{(s;\,O)} e^{-i(X\cdot U)}\,d\,V_X\right\}e^{-i(Y\cdot U)}\,M^{(m)}(r;\,U)\,d\,V_U. \tag{8. 27}$$

Now since

$$\lim_{r\to 0}\frac{1}{[v(r)]^m}\left\{\int_{(r;\,O)} e^{i(X\cdot T)}\,d\,V_T\right\}^m = 1, \tag{8. 28}$$

and since $f(X)$ is summable, it follows that

$$\int_{(s;\,Y)} f(X)\,d\,V_X = \lim_{r\to 0}\frac{1}{(2\,\pi)^{n/2}}\frac{1}{[v(r)]^m}\int_\infty\left\{\int_{(s;\,O)} e^{-i(X\cdot U)}\,d\,V_X\right\}e^{-i(Y\cdot U)}\,M^{(m)}(r;\,U)\,d\,V_U. \tag{8. 29}$$

The right hand side is precisely an mth order Stieltjes integral. We have, then,

$$\int_{(s;\,Y)} f(X)\,d\,V_X = \frac{1}{(2\,\pi)^{n/2}}\int_\infty\left\{\int_{(s;\,O)} e^{-i(X\cdot U)}\,d\,V_X\right\}e^{-i(Y\cdot U)}\,d_U^m\,M^{(m)}, \tag{8. 30}$$

and, therefore, for almost every X,

$$f(X) = \lim_{s\to 0}\frac{1}{(2\,\pi)^{n/2}}\frac{1}{v(r)}\int_\infty\left\{\int_{(s;\,O)} e^{-i(T\cdot U)}\,d\,V_T\right\}e^{-i(X\cdot U)}\,d_U^m\,M^{(m)}. \tag{8. 31}$$

That the function $M^{(m)}(r;\,U)$ constitutes a harmonic analysis of the given $f(X)$ is established by the following considerations. It is a matter of simple calculation to show that if $f(X)$ is itself q. s. over ∞, $M^{(1)}(r;\,U)$ exists and has for its value the total mass in the sphere $(r;\,U)$ of a distribution of density $g(U)$. It is similarly easy to show that if $f(X)$ is possessed of a Fourier series development, $M^{(1)}(r;\,U)$ exists as a limit in the mean and is equal to the sum of those Fourier coefficients a_k which correspond to frequency points U lying in $(r;\,U)$. As we have seen above, $M^{(1)}(r;\,U)$ determines the mass distributions which constitute the harmonic analysis of $f(X)$. In the same fashion $M^{(2)}(r;\,U)$ determines $M^{(1)}(r;\,U)$; etc.

While, then, we are justified in considering that $M^{(m)}(r;\,U)$ is a harmonic analysis of $f(X)$, we must yet determine for what class of functions $f(X)$ the region-function M will exist at least for some integer m. It can be shown that the function

$$\int_{(r;\,O)} e^{i(X\cdot T)}\,d\,V_T \tag{8.32}$$

is bounded for all X and is of the order of $|X|^{-\frac{n+1}{2}}$ for large values of $|X|$. This at once establishes the fact that if, for some positive p, the quotient

$$\frac{f(X)}{1+|X|^r} \tag{8.33}$$

is q. s. over ∞, then a value of m can be determined for which $M^{(m)}$ will exist. Furthermore, a generalization to n dimensions of a Tauberian theorem such as given by Jacob readily shows that if $f(X)$ is q. s. over all finite regions and is such that for some positive p the expression

$$\frac{1}{[v(r)]^p}\int_{(r;\,O)} |f(X)|^2\,d\,V_X \tag{8.34}$$

is bounded for sufficiently large values of r, then it too belongs to the class in question. An n-tuply periodic function is included in this last type of function $f(X)$. Essentially, then, the functions which we can harmonically analyze are those which are »algebraic on the average» as $|X|\to\infty$.

When we come to the study of the energy spectrum of a given function $f(X)$ we again subject the function to a harmonic analysis but not in such great detail as before. In the corresponding mass spread we are no longer interested in the complex phase of the various masses and densities present, but solely in their absolute values, or, more precisely, in the squares of these absolute values. Obviously, again, there will be different orders of density, or, as we may say, different orders of energy in the component oscillations. For the functions of the class described above there will exist for some value of $k\geq 0$ the limit

$$\varphi(X)=\lim_{r\to\infty}\frac{1}{[v(r)]^k}\int_{(r;\,O)} f(T)\bar{f}(T-X)\,d\,V_T. \tag{8.35}$$

The harmonic analysis of this latter function will show that its mass distribution consists of the squares of the absolute values of the highest order densities that appear in the distribution corresponding to $f(X)$. Terms of lower energy level do not appear. Thus, the total mass of the distribution corresponding to $\varphi(X)$ does not necessarily coincide with although it will never exceed the total energy which could be calculated from the spread associated with $f(X)$.

Precisely as in the one-dimensional case it can be shown that

$$|\varphi(X)| \leq \varphi(0) = \lim_{r \to \infty} \frac{1}{[v(r)]^k} \int\limits_{(r;\,O)} |f(T)|^2 \, d\,V_T. \tag{8.36}$$

From this boundedness it follows that the product

$$\varphi(X) \left\{ \int\limits_{(r;\,O)} e^{i(X \cdot T)} \, d\,V_T \right\} \tag{8.37}$$

is q. s. over ∞, and hence that there exists $M^{(1)}(r;\,U)$ formed with respect to $\varphi(X)$. This function is at present only defined as a limit in the mean. It is desired to show that it can be so defined that for any given U it will be a monotonic non-decreasing function of r. This is carried out as when $n=1$ by a series of limiting-processes which do not alter any monotonic properties. To begin with one shows that if

$$\left.\begin{aligned} \varphi_s(X) &= \frac{1}{[v(s)]^k} \int\limits_{\infty} f_s(T) \bar{f}_s(T - X) \, d\,V_T; \\ f_s(T) &= \begin{cases} f(T); & [T \text{ in } (s;\,0)] \\ 0\;; & [T \text{ elsewhere}] \end{cases} \end{aligned}\right\} \tag{8.38}$$

then

$$\varphi(X) = \lim_{s \to \infty} \varphi_s(X). \tag{8.39}$$

Furthermore one notes that

$$\int\limits_{\infty} \varphi_s(X) \, e^{i(X \cdot Y)} \, d\,V_X = \frac{1}{[v(s)]^k} \left| \int\limits_{\infty} f_s(T) \, e^{i(T \cdot Y)} \, d\,V_T \right|^2 \geq 0, \tag{8.40}$$

precisely as in the one-dimensional case. From this it follows that the function

$$M_s(r;\,U) = \frac{1}{(2\pi)^n}\int_{\infty} \varphi_s(X)\left\{\int_{(r;\,O)} e^{i(X\cdot T)}\,dV_T\right\} e^{i(X\cdot U)}\,dV_X$$

$$= \frac{1}{(2\pi)^n}\int_{(r;\,U)} dV_T \int_{\infty} \varphi_s(X)\,e^{i(X\cdot T)}\,dV_X \tag{8.41}$$

for given U is monotone non-decreasing in r. It is readily seen that

$$\int_{(r';\,U')} M(r;\,U)\,dV_U = \lim_{S\to\infty}\int_{(r';\,U')} M_s(r;\,U)\,dV_U, \tag{8.42}$$

and hence that for given r' and U' the integral on the left has the same monotone property with respect to r. Finally since, for almost every U',

$$M(r;\,U') = \lim_{r'\to 0}\frac{1}{v(r')}\int_{(r';\,U')} M(r;\,U)\,dV_U, \tag{8.43}$$

we see that, in so far as it is thus defined, $M(r;\,U)$ has the desired property. Since the point-set on which M is not defined constitutes at most a set of zero measure, it is a simple matter to define M for all r and U so that for each U will be a monotonic non-decreasing function of r.

The total spectral intensity of $f(X)$, or more accurately, the total spectral intensity of those components which are associated with the maximum energy level, is given by the limit

$$\lim_{r\to\infty} M(r;\,U), \tag{8.44}$$

if this exists. One fairly readily shows that

$$M(r;\,U) \le \varphi(0), \tag{8.45}$$

and, because of the monotony, hence that the limit in question exists and lies between 0 and $\varphi(0)$. The details of the argument whereby one next shows that

$$\lim_{r\to\infty} M(r;\,U) = \lim_{r'\to 0}\frac{1}{v(r')}\int_{(r';\,O)} \varphi(X)\,dV_X \le \varphi(0) \tag{8.46}$$

will be omitted here to avoid too much complication.

9. Coherency matrices.

The spectrum theory of our earlier sections is a theory of the spectrum of an individual function. There are, however, many phenomena intimately connected with harmonic analysis which refer to several functions considered simultaneously. Chief among these are the phenomena of coherency and incoherency, of interference, and of polarization.

It is known to every beginner in physics that two rays of light from the same source may interfere: that is, they may be superimposed to form a darkness, or else a light more intense than is ordinarily formed by two rays of light of their respective intensities. On the other hand, two rays of light from independent sources or from different parts of the same source never exhibit this phenomenon. The former rays are said to be *coherent*, and the latter to be *incoherent*. Although it is mathematically impossible for two truly sinusoidal oscillations to be incoherent, even the most purely monochromatic light which we can sensibly produce never coheres with similar light from another source.

The physicist's explanation of incoherency is the following: the interference pattern produced by two sources of light depends on their relative displacement in phase. Now, the relative phase of two sensibly monochromatic sources of light is able to assume all possible values, and since light probably consists in a series of approximately sinusoidal trains of oscillations each lasting but a small portion of a millionth of a second, this relative phase assumes in any sensible interval all possible values with a uniform distribution which averages out light and dark bands into a sensibly uniform illumination.

This explanation of incoherency is unquestionable adequate to account for the phenomenon which it was invented to explain. Nevertheless, it is desirable to have a theory of coherency and incoherency which does not postulate a hypothetical set of constituent harmonic trains of oscillations, which at any rate must become merged in the general electromagnetic oscillation constituting the light. The present section is devoted to the development of a theory of coherency which is as direct as the theory of this paper concerning the harmonic analysis of a single function, and indeed forms a natural extension of the latter.

In interference, the components of the electromagnetic field of the constituent light rays combine additively. Accordingly, the theory of coherency and interference must be a theory of the harmonic analysis of all functions which can be obtained from a given set by linear combination. Let us see what the outlines of this theory are.

We start from a class of functions, $f_k(t)$, in general complex, and defined for all real arguments between $-\infty$ and ∞. For the present we shall assume this class of functions to be finite, although this restriction is not essential. Let

$$f(t)=a_1 f_1(t)+a_2 f_2(t)+\cdots+a_n f_n(t) \tag{9.01}$$

be the general linear combination of functions of the set. We shall have

$$\varphi(\tau)=\lim_{T\to\infty}\frac{1}{2T}\int_{-T}^{T} f(t+\tau)\bar{f}(t)\,dt=\sum_{j,k=1}^{n}\sum a_j\bar{a}_k\lim_{T\to\infty}\frac{1}{2T}\int_{-T}^{T} f_j(t+\tau)\bar{f}_k(t)\,dt \tag{9.02}$$

in case the latter limits exist. The necessary and sufficient condition for this to exist for all linear combinations $f(t)$ of functions of the set is that

$$\varphi_{jk}(\tau)=\lim_{T\to\infty}\frac{1}{2T}\int_{-T}^{T} f_j(t+\tau)\bar{f}_k(t)\,dt \tag{9.03}$$

should exist for every j, k, and τ. Then

$$\varphi(\tau)=\sum_j\sum_k a_j\bar{a}_k\varphi_{jk}(\tau). \tag{9.04}$$

Again, we shall have formally

$$S(u)=\frac{1}{2\pi}\int_{-\infty}^{\infty}\varphi(\tau)\frac{e^{iu\tau}-1}{i\tau}\,d\tau=\sum_j\sum_k a_j\bar{a}_k\frac{1}{2\pi}\int_{-\infty}^{\infty}\varphi_{jk}(\tau)\frac{e^{iu\tau}-1}{i\tau}\,d\tau, \tag{9.05}$$

where we may write

$$S_{jk}(u)=\frac{1}{2\pi}\int_{-\infty}^{\infty}\varphi_{jk}(\tau)\frac{e^{iu\tau}-1}{i\tau}\,d\tau. \tag{9.06}$$

If $\varphi_{jk}(\tau)$ exists for every j, k, and τ, we may readily show that each $S_{jk}(u)$ exists. Clearly

$$\varphi_{kj}(\tau)=\lim_{T\to\infty}\frac{1}{2T}\int_{-T}^{T} f_k(t+\tau)\bar{f}_j(t)\,dt=\lim_{T\to\infty}\frac{1}{2T}\int_{-T}^{T}\bar{f}_j(t-\tau)f_k(t)\,dt=\bar{\varphi}_{jk}(-\tau) \tag{9.07}$$

and

$$S_{kj}(u) = \frac{1}{2\pi}\int_{-\infty}^{\infty} \bar{\varphi}_{jk}(-\tau)\frac{e^{iu\tau}-1}{i\tau}\,d\tau = \frac{1}{2\pi}\int_{-\infty}^{\infty} \bar{\varphi}_{jk}(\tau)\frac{e^{-iu\tau}-1}{-i\tau}\,d\tau = \bar{S}_{jk}(u), \tag{9.08}$$

so that the matrix

$$\| S_{jk}(u) \|$$

is Hermitian. This matrix determines the spectra of all possible linear combinations of $f_1(t), \ldots, f_n(t)$. Since it determines the precise coherency relations of the functions in question, we shall call it the coherency matrix.

Let us subject the functions $f_k(t)$ to the linear transformation

$$g_j(t) = \sum_k a_{jk} f_k(t). \tag{9.09}$$

Then

$$\psi_{jk}(\tau) = \lim_{T\to\infty}\frac{1}{2T}\int_{-T}^{T} g_j(t+\tau)\,\bar{g}_k(t)\,dt = \sum_l \sum_m a_{jl}\bar{a}_{km}\,\varphi_{lm}(\tau) \tag{9.10}$$

and

$$T_{jk}(u) = \frac{1}{2\pi}\int_{-\infty}^{\infty} \psi_{jk}(\tau)\frac{e^{iu\tau}-1}{i\tau}\,d\tau = \sum_l \sum_m a_{jl}\bar{a}_{km}\,S_{lm}(u). \tag{9.11}$$

Thus the new coherency matrix $\| T_{jk}(u) \|$ may be written

$$M_1 \cdot \| S_{jk}(u) \| \cdot M_2, \tag{9.12}$$

where

$$M_1 = \| a_{jk} \|, \tag{9.13}$$

and

$$M_2 = \| \bar{a}_{kj} \|. \tag{9.14}$$

In case the transformation with matrix M_1 has the property

$$M_1 \cdot M_2 = M_2 \cdot M_1 = 1, \tag{9.15}$$

it is said to be *unitary*. For such transformations,

$$\| T_{jk} \| = M_1 \cdot \| S_{jk} \| \cdot M_1^{-1}. \tag{9.16}$$

A matrix $\| a_{jk} \|$ is said to be in diagonal form if all the terms a_{jk} for which $j \neq k$ are identically zero. By a theorem of Weyl[5], every Hermitian matrix may be transformed into a diagonal matrix by a unitary transformation. Since we may regard a diagonal matrix as representing a set of completely incoherent phenomena, this transformation is of fundamental importance in the characterization of the state of coherency of the functions determining the matrix. Together with the numerical values of the diagonal elements of the diagonal matrix, it indeed consitutes a complete characterization of the state of coherency of the original function. In the case the values of the diagonal elements are distinct, this characterization is indeed to be carried through in but a single way.

The production of incoherent functions is a simple matter, when we have once settled the existence theory of functions with given types of spectra. Let $f_1(t)$ be any bounded function such that $\varphi(\tau)$ and consequently $S(u)$ exists. Then if

$$f_2(t) = f_1(t)\, e^{i\sqrt{|t|}}, \tag{9.17}$$

we have

$$\lim_{T\to\infty} \frac{1}{2T} \int_{-T}^{T} f_2(t+\tau) \bar{f}_2(t)\, dt = \lim_{T\to\infty} \frac{1}{2T} \int_{-T}^{T} f_1(t+\tau) \bar{f}_1(t)\, e^{i(\sqrt{|t+\tau|}-\sqrt{|t|})}\, dt$$

$$= \lim_{T\to\infty} \frac{1}{2T} \int_{-T}^{T} f_1(t+\tau) \bar{f}_1(t) \exp\left(i \frac{|t+\tau| - |t|}{\sqrt{|t+\tau|} + \sqrt{|t|}}\right) dt, \tag{9.18}$$

and hence, since $\lim\limits_{t\to\pm\infty} \exp\left(i \frac{|t+\tau| - |t|}{\sqrt{|t+\tau|} + \sqrt{|t|}}\right) = 1$,

$$\varphi_{22}(\tau) = \varphi_{11}(\tau) \tag{9.19}$$

and

$$S_{22}(u) = S_{11}(u). \tag{9.20}$$

On the other hand, if for example $f(t) = e^{i\lambda t}$,

$$\left.\begin{aligned} \varphi_{12}(\tau) &= \lim_{T\to\infty} \frac{1}{2T} \int_{-T}^{T} f_1(t+\tau) \bar{f}_1(t)\, e^{-i\sqrt{|t|}}\, dt \equiv 0; \\ S_{12}(u) &= S_{21}(u) = 0. \end{aligned}\right\} \tag{9.21}$$

Thus the coherency matrix of $f_1(t)$ and $f_2(t)$ is

$$S_{11}(u)\left\|\begin{matrix} 1 & 0 \\ 0 & 1 \end{matrix}\right\|. \tag{9. 22}$$

The coherency matrix of $\sqrt{2}\, f_1(t)$ and 0 is

$$S_{11}(u)\left\|\begin{matrix} 2 & 0 \\ 0 & 0 \end{matrix}\right\|; \tag{9. 23}$$

that of $f_1(t)$ and $f_1(t)$ is

$$S_{11}(u)\left\|\begin{matrix} 1 & 1 \\ 1 & 1 \end{matrix}\right\|; \tag{9. 24}$$

that of $f_1(t)$ and $i f_1(t)$ is

$$S_{11}(u)\left\|\begin{matrix} 1 & -i \\ i & 1 \end{matrix}\right\|. \tag{9. 25}$$

Let it be noted that the coherency matrices of real functions are in general complex. Thus if

$$f_2(t) = f_1(t+\lambda), \tag{9. 26}$$

we have

$$\left.\begin{aligned} \varphi_{22}(\tau) &= \varphi_{11}(\tau); \\ \varphi_{12}(\tau) &= \lim_{T\to\infty} \frac{1}{2T}\int_{-T}^{T} f_1(t+\lambda+\tau) f_1(t)\, dt \\ &= \varphi_{11}(\tau+\lambda); \end{aligned}\right\} \tag{9. 27}$$

and hence

$$\left.\begin{aligned} &S_{22}(u) = S_{11}(u); \\ &S_{12}(u) = \int_{-\infty}^{\infty} \varphi_{11}(\tau+\lambda)\frac{e^{iu\tau}-1}{i\tau}\, d\tau = \int_{-\infty}^{u} e^{-iv\lambda}\, d\, S_{11}(v) + S_{12}(-\infty) \end{aligned}\right\} \tag{9. 28}$$

giving the coherency matrix with derivative

$$S'_{11}(u)\left\|\begin{matrix} 1 & e^{-iv\lambda} \\ e^{iv\lambda} & 1 \end{matrix}\right\|. \tag{9. 29}$$

In optics, coherency is generally considered for light of one particular frequency. From that standpoint, the coherency of a set of functions $f_1(t), \ldots, f_n(t)$

with continuous differentiable spectra may be regarded as determined for frequency u by the matrix

$$\left\| \begin{matrix} S^1_{11}(u), & \ldots, & S^1_{1n}(u) \\ \cdot & \cdot & \cdot \\ \cdot & \cdot & \cdot \\ S^1_{n1}(u), & \ldots, & S^1_{nn}(u) \end{matrix} \right\|; \tag{9.30}$$

or in the case of functions with line spectra, by

$$\left\| \begin{matrix} S_{11}(u+0)-S_{11}(u-0), & \ldots, & S_{1n}(u+0)-S_{1n}(u-0) \\ \cdot & \cdot & \cdot \\ \cdot & \cdot & \cdot \\ S_{n1}(u+0)-S_{n1}(u-0), & \ldots, & S_{nn}(u+0)-S_{nn}(u-0) \end{matrix} \right\|. \tag{9.31}$$

We may regard these matrices in a secondary sense as coherency matrices.

Coherency matrices of two functions are of particular interest in connection with the characterization of the state of polarization of light. As everyone knows, this characterization is identically the characterization of the state of coherency between two components of the electric vector at right angles to one another. With this interpretation, matrix (9.22) represents unpolarized light, matrix (9.23) light polarized completely in one plane, while

$$\left\| \begin{matrix} 0 & 0 \\ 0 & 2 \end{matrix} \right\| \tag{9.32}$$

represents light completely polarized in a plane perpendicular to the first. Matrix (9.24) and matrix

$$\left\| \begin{matrix} 1 & -1 \\ -1 & 1 \end{matrix} \right\| \tag{9.33}$$

represent light polarized respectively at 45° and at 135° to the first direction. Matrix (9.25) and matrix

$$\left\| \begin{matrix} 1 & i \\ -i & 1 \end{matrix} \right\| \tag{9.34}$$

represent respectively light polarized circularly in a counter-clockwise and a clockwise direction.

When the matrix of completely polarized light, whether linearly, elliptically, or circularly polarized, is brought into diagonal form by a linear unitary transformation, the resulting diagonal matrix will have only one element distinct from 0. On the other hand, completely unpolarized light has the diagonal terms equal. This suggests as a measure of the amount of polarization of the diagonal matrix

$$\begin{Vmatrix} a & 0 \\ 0 & b \end{Vmatrix}, \tag{9. 35}$$

or of any other matrix equivalent to it under a unitary transformation, the quantity

$$a - b. \tag{9. 36}$$

If we subtract from our original diagonal matrix the completely incoherent matrix

$$\begin{Vmatrix} \frac{a+b}{2} & 0 \\ 0 & \frac{a+b}{2} \end{Vmatrix}, \tag{9. 37}$$

which is invariant under every unitary transformation, we get the matrix

$$\begin{Vmatrix} \frac{a-b}{2} & 0 \\ 0 & \frac{b-a}{2} \end{Vmatrix}, \tag{9. 38}$$

which may be regarded as a representative of the quantity $a-b$. This suggests that given any coherency matrix

$$\begin{Vmatrix} A & B+Ci \\ B+Ci & D \end{Vmatrix}, \tag{9. 39}$$

we may take $A+D$ to represent the intensity of the corresponding light, and the matrix

$$\begin{Vmatrix} \frac{A-D}{2} & B-Ci \\ B+Ci & \frac{D-A}{2} \end{Vmatrix} \tag{9. 40}$$

as its polarization. Thus horizontal polarization is represented by the matrix

$$\begin{Vmatrix} 1 & 0 \\ 0 & -1 \end{Vmatrix}; \tag{9.41}$$

polarization at 45° by the matrix

$$\begin{Vmatrix} 0 & 1 \\ 1 & 0 \end{Vmatrix}; \tag{9.42}$$

and circular polarization by the matrix

$$\begin{Vmatrix} 0 & i \\ -i & 0 \end{Vmatrix}. \tag{9.43}$$

These are the same matrices which Jordan, Dirac and Weyl have employed to such advantage in the theory of quanta.

Since the most general Hermitian matrix of the second order may be written

$$\begin{Vmatrix} \alpha+\beta & \gamma+\delta i \\ \gamma-\delta i & \alpha-\beta \end{Vmatrix} = \alpha \begin{Vmatrix} 1 & 0 \\ 0 & 1 \end{Vmatrix} + \beta \begin{Vmatrix} 1 & 0 \\ 0 & -1 \end{Vmatrix} + \gamma \begin{Vmatrix} 0 & 1 \\ 1 & 0 \end{Vmatrix} + \delta \begin{Vmatrix} 0 & i \\ -i & 0 \end{Vmatrix}, \tag{9.44}$$

it appears that all light may characterized as to its state of polarization by given the total amount of light it contains, the excess of the amount polarized at 0° over that polarized at 90°, the excess of the amount polarized at 45° over that polarized at 135° and the excess of that polarized circularly to the right over that polarized circularly to the left. This characterization is complete and univocal. The total intensity of the light may be read off any sort of a photometer. The excess of light polarized horizontally over that polarized vertically may be determined by a doubly refracting crystal in one orientation, and the excess of light polarized at 45° over that polarized at 135° by the same crystal in another orientation. It is possible furthermore to devise an instrument which will read off the amount of circular polarization in the light in question. The three latter instruments possess some very remarkable group properties with respect to one another. Either portion of the light emerging from any one of the instruments will behave towards the other two exactly like unpolarized light. Rotation of the plane of polarization of the light through 45° will change the reading of the first of the last three instruments into that of the second, and the reading of the second into minus that of the first, leaving the reading of the third unchanged. There are precisely analogous unitary transformations interchanging any other pair of the three instruments, leaving the reading of the remaining instrument untouched. These transformations together with their powers and the identical transformation form a group.

A fact concerning polarized light which is so apparently obvious that it is generally regarded as not needing any proof is that all light is a case or limiting case of partially elliptically polarized light. It is nevertheless desirable to prove this statement. Completely elliptically polarized light with the coordinate axes as principal axes has a coherency matrix of the form

$$\begin{Vmatrix} A^2 & -ABi \\ ABi & B^2 \end{Vmatrix}; \tag{9.45}$$

and hence partially polarized light with the same principal axes has a coherency matrix of the form

$$P = \begin{Vmatrix} A^2+D^2 & -ABi \\ ABi & B^2+D^2 \end{Vmatrix}. \tag{9.46}$$

We wish to show that the general coherency matrix

$$M = \begin{Vmatrix} \alpha & \gamma-\delta i \\ \gamma+\delta i & \beta \end{Vmatrix} \tag{9.47}$$

may be transformed into this form by a real unitary transformation in such a way that

$$T \,.\, M \,.\, T^{-1} = P.$$

To do this, we need only put

$$T = \begin{Vmatrix} \cos\varphi & \sin\varphi \\ \sin\varphi & -\cos\varphi \end{Vmatrix}; \text{ where } \tan 2\varphi = \frac{2\gamma}{\alpha-\beta}. \tag{9.48}$$

Thus the axes of polarization of M are 1 and 2 directions when we replace $f_1(t)$ and $f_2(t)$ by

$$\left.\begin{aligned} g_1(t) &= f_1(t)\cos\varphi + f_2(t)\sin\varphi; \\ g_2(t) &= -f_2(t)\cos\varphi + f_1(t)\sin\varphi; \end{aligned}\right\} \tag{9.49}$$

the »lengths» of these axes are respectively

$$\left.\begin{aligned} A &= \left\{\frac{1}{2}\,[(\alpha-\beta)^2+4\,(\gamma^2+\delta^2)]^{1/2}+\frac{1}{2}\,[(\alpha-\beta)^2+4\,\gamma^2]^{1/2}\right\}^{1/2};\\ B &= \left\{\frac{1}{2}\,[(\alpha-\beta)^2+4\,(\gamma^2+\delta^2)]^{1/2}-\frac{1}{2}\,[(\alpha-\beta)^2+4\,\gamma^2]^{1/2}\right\}^{1/2}; \end{aligned}\right\} \tag{9. 50}$$

and the percentage of polarization

$$100\left(1-\frac{4\,(\gamma^2+\delta^2)}{(\alpha+\beta)^2}\right)^{1/2}. \tag{9. 51}$$

The connection between coherency matrices and optical instruments, which we have already mentioned in the case of polarized light, is of far more general applicability. An optical instrument is a method, linear in electric and magnetic field vectors, of transforming a light input into a light output. In general, this transformation, in the language of Volterra[5], belongs to the group of the closed cycle with respect to the time, in the sense that it is independent of the position of our initial instant in time. Such a transformation leaves a simple harmonic input still simply harmonic in the time, although in general with a shift in phase.

An example of an optical instrument is a microscope. This may be regarded as a means of making an electromagnetic disturbance in the image plane depend linearly on a given electromagnetic disturbance in the object plane. Telescopes, spectroscopes, Nicol prisms, etc., serve as further examples of optical instruments in the sense in question. Among these, a particularly interesting ideal type is the conservative optical instrument, in which the power of the input and the power of the output are identical. This power depends quadratically on the electric and magnetic vectors, so that a conservative optical instrument has a quadratic invariant for the corresponding transformations. When only terms of the one frequency of e^{iut} are considered, this quadratic positive invariant becomes Hermitian, and has essentially the same properties as the expression

$$x_1\bar{x}_1+x_2\bar{x}_2+\cdots+x_n\bar{x}_n \tag{9. 52}$$

which is invariant under all unitary transformations of $x_1, \ldots, x_n$. Thus the theory of the group of unitary transformations is physically applicable, not only in quantum mechanics, where Weyl has already employed it so successfully, but even in classical optics. It is the conviction of the author that this analogy is not merely an accident, but is due to a deep-lying connection between the two theories.

In quantum mechanics, while all the terms of a matrix enter in an essential way into its transformation theory, only diagonal terms are given an immediate physical significance. This is also in precise accord with the optical situation. Every optical observation ends with the measurement of an energy or power, either by direct bolometric or thermometric means, or by the observations of a visual intensity or the blackening of a photographic plate. Every such observation means the more or less complete determination of some diagonal term. The non-diagonal terms of a coherency matrix of light only have signifiance in so far as they enable us to predict the energies or intensities which the light will show after having been subjected to some linear transformation or optical instrument. This fact that new diagonal terms after a transformation cannot be read off from the old diagonal terms before a transformation, without the intervention of non-diagonal terms, is the optical analogue for the principle of indetermination in quantum theory, according to which observations on the momentum of an electron alone cannot yield a single value if its position is known, and vice versa. The statement that every observation of an electron affects its properties has the following analogy: if two optical instruments are arranged in series, the taking of a reading from the first will involve the interposition of a ground-glass screen or photographic plate between the two, and such a plate will destroy the phase relations of the coherency matrix of the emitted light, replacing it by the diagonal matrix with the same diagonal terms. Thus the observation of the output of the first instrument alters the output of the second. In this case, the possibility of taking part of the output of the first instrument for reading by a thinly silvered mirror warns us not to try to push the analogy with quantum theory too far.

Coherency matrices form a close analogue to the correlation matrices long familiar in statistical theory. If we have a set of n observations $x^{(1)}, x^{(2)}, \ldots, x^{(n)}$ all made together, and this set of observations is repeated on the occasions 1, 2, ..., m thus yielding sets $x_1^{(1)}, \ldots x_1^{(n)}$; $x_2^{(1)}, \ldots, x_2^{(n)}$; ...; ...; $x_m^{(1)}, \ldots, x_m^{(n)}$, the correlation matrix of these observations is

$$\left\| \begin{array}{cccc} \sum_1^m (x_k^{(1)})^2, & \sum_1^m x_k^{(1)} x_k^{(2)}, & \cdots\cdots, & \sum_1^m x_k^{(1)} x_k^{(n)} \\ \sum_1^m x_k^{(2)} x_k^{(1)}, & \sum_1^m (x_k^{(2)})^2, & \cdots\cdots, & \sum_1^m x_k^{(2)} x_k^{(n)} \\ \cdots & \cdots & \cdots & \cdots \\ \cdots & \cdots & \cdots & \cdots \\ \sum_1^m x_k^{(n)} x_k^{(1)}, & \sum_1^m x_k^{(n)} x_k^{(2)}, & \cdots\cdots, & \sum_1^m (x_k^{(n)})^2 \end{array} \right\| . \tag{9.53}$$

This symmetrical matrix represents the entire amount and kind of linear relationship to be observed between the different observations in question. The further analysis of the information yielded by a correlation matrix depends on the nature of the data to be analysed. Thus if the two observations of each set are the x and y coordinates of the position of a shot on a target, the rotations of the x and y axes have a concrete geometrical meaning, and the question of reducing the matrix to diagonal form by a rotation of axes is the significant question of determining the ellipse which best represents the distribution of holes in the target. On the other hand, if the quantities whose correlation we are investigating are the price of wheat x in dollars per bushel and the marriage rate y, rotations have no significance, as there is no common scale, while on the other hand, the significant information yielded by the matrix must be invariant under the transformations

$$\left. \begin{aligned} x_1 &= kx; \\ y_1 &= ly. \end{aligned} \right\} \tag{9.54}$$

The so-called coefficients of correlation and of partial correlation and the lines of regression of x on y and of y on x have this type of invariance.

Correlation matrices and their derived quantities are the tool for the statistical analysis of what is known as frequency series, series of data where no such variable as the time enters as a parameter. In the study of meteorology, of business cycles, and of many other phenomena of interest to the statistician, on the other hand, we must discuss time series, where the relations of the data in time are essential. The proper analysis of these has long been a moot point among statisticians and economists. As far as it is linear relationships which we are seeking for, it is only reasonable to suppose that coherency matrices must play the same rôle for time series which correlation matrices play for frequency series. In statistical work, the group of transformations which will most frequently be permissible is as before

$$\begin{aligned} g_1(t) &= A f_1(t); \\ g_2(t) &= B f_2(t). \end{aligned} \qquad [A \text{ and } B \text{ real}] \tag{9.55}$$

Under this group, the significant invariants of the Hermitian matrix

$$\begin{Vmatrix} S^1_{11}(u) & S^1_{12}(u) \\ S^1_{21}(u) & S^1_{22}(u) \end{Vmatrix} \tag{9.56}$$

are

$$r(u) = S^1_{12}(u) \left[S^1_{11}(u)\, S^1_{22}(u) \right]^{-1/2}; \tag{9.57}$$

which we may call the coefficient of coherency of f_1 and f_2 for frequency u, and

$$\sigma_1(u) = \frac{S^1_{12}(u) \sqrt{S^1_{11}(u)}}{S^1_{22}(u)} \text{ and } \sigma_2(u) = \frac{S^1_{21}(u) \sqrt{S^1_{22}(u)}}{S^1_{11}(u)}, \tag{9.58}$$

the coefficients of regression respectively of f_1 on f_2 and of f_2 on f_1. The modulus of the coefficient of coherency represents the amount of linear coherency between $f_1(t)$ and $f_2(t)$ and the argument the phase-lag of this coherency. The coefficients of regression determine in addition the relative scale for equivalent changes of f_1 and f_2.

The computation of coefficients of coherency and of regression is to be done in the steps indicated in their definition. In the case where only a finite set of real data are at our disposal, distributed at equal unit intervals from 0 to n, say $x_0, x_1, \ldots, x_n$ and $y_0, y_1, \ldots, y_n$, the steps of our computation are:

$$\left.\begin{aligned} (\varphi_k)_{11} &= \frac{1}{n} \sum_0^{n-k} x_j\, x_{j+k}; \\ (\varphi_k)_{12} &= \frac{1}{n} \sum_0^{n-k} x_j\, y_{j+k}; \\ (\varphi_k)_{21} &= \frac{1}{n} \sum_0^{n-k} y_j\, x_{j+k}; \\ (\varphi_k)_{22} &= \frac{1}{n} \sum_0^{n-k} y_i\, y_{j+k}; \end{aligned}\right\} \qquad [0 \leq k \leq n] \tag{9.59}$$

$$r(u)=\frac{\sum_{0}^{n}\{[(\varphi_k)_{21}+(\varphi_k)_{12}]\cos ku+i\,[(\varphi_k)_{21}-(\varphi_k)_{12}]\sin ku\}-(\varphi_0)_{12}/2}{2\left[\sum_{0}^{n}(\varphi_k)_{11}\cos ku-(\varphi_0)_{11}/2\right]^{1/2}\left[\sum_{0}^{n}(\varphi_k)_{22}\cos ku-(\varphi_0)_{22}/2\right]^{1/2}}; \tag{9.60}$$

$$\left.\begin{aligned}\sigma_1(u)&=r(u)\frac{\left[\sum_{0}^{n}(\varphi_k)_{11}\cos ku-(\varphi_0)_{11}/2\right]^{1/2}}{\left[\sum_{0}^{n}(\varphi_k)_{22}\cos ku-(\varphi_0)_{22}/2\right]^{1/2}};\\ \sigma_2(u)&=\bar{r}(u)\frac{\left[\sum_{0}^{n}(\varphi_k)_{22}\cos ku-(\varphi_0)_{22}/2\right]^{1/2}}{\left[\sum_{0}^{n}(\varphi_k)_{11}\cos ku-(\varphi_0)_{11}/2\right]^{1/2}}.\end{aligned}\right\} \tag{9.61}$$

In case we have at our disposal methods for performing the Fourier transformation, we may compute these coefficients directly from graphs. Several devices for this purpose are now being developed in the laboratory of Professor V. Bush in the Department of Electrical Engineering of the Massachusetts Institut of Technology.

10. Harmonic analysis and transformation groups.

Inasmuch, as the theory of Fourier series forms a special chapter in the theory of expansions in general sets of normal and orthogonal functions, it is reasonable to expect that the theory developed in the present paper is but a special chapter in the theory of general orthogonal developments. An attempt to translate the present theory into more general terms, however, incurs at once somewhat serious difficulties. This is due to the fact that the theory of the Fourier series involves only one fundamental Hermitian form,

$$\frac{1}{2\pi}\int_{-\pi}^{\pi} f(x)\bar{f}(x)\,dx; \tag{10.01}$$

the closely related theory of the Fourier integral involves only the analogous form

$$\int_{-\infty}^{\infty} f(x)\bar{f}(x)\,dx; \tag{10.02}$$

while the present paper involves besides this latter form the singular quadratic form

$$M_x(f(x)) = \lim_{T\to\infty} \frac{1}{2T}\int_{-T}^{T} f(x)\bar{f}(x)\,dx. \tag{10.03}$$

The forms (10.02) and (10.03) are quite independent of one another in their formal properties, but the complex exponentials e^{iux} stand in close relation to both of them: to (10.02) because if $a < b$, $c < d$,

$$\int_{-\infty}^{\infty} dx \int_a^b e^{iux}\,du \int_c^d e^{-ivx}\,dv = 2\pi \text{ [length common to } (a, b) \text{ and } (c, d)]; \tag{10.04}$$

and to (10.03) because

$$\lim_{T\to\infty} \frac{1}{2T}\int_{-T}^{T} e^{iux}\,e^{-ivx}\,dx = \begin{cases} 0; & [u \neq v] \\ 1 & [u = v]. \end{cases} \tag{10.05}$$

In the classical Plancherel theory, only the first form is in evidence; in the Bohr theory of almost periodic functions, only the second; in the theory of the present paper, the two play an equal rôle.

Weyl has developed in some detail the relations between the theory of unitary groups and the theory of periodic and almost periodic functions. The groups which he introduces are one parameter groups of linear functional transformations leaving (10.03) invariant. The Weyl theory is manifestly not susceptible to an extension to more general forms of harmonic analysis, unless a way is found to take cognizance of the invariance of (10.02) as well. This is the purpose of the present section.

Let us restrict the functions $f(x)$ which we discuss in the present section to those for which

$$\int_{-\infty}^{\infty} \frac{|f(x)|^2}{1 + x^2}\,dx \tag{10.06}$$

is finite. Let

$$f_A(x)=\begin{cases} 0; & [|x|>A] \\ f(x); & [|x|\leq A] \end{cases} \tag{10.07}$$

and let s_A be the transformation leading from $f(x)$ to $f_A(x)$. Let T be a transformation which is linear, and with an inverse, and is defined for all functions $f(x)$ subject to the finiteness of (10.06). Let T preserve (10.02) invariant, and in case (10.03) is finite, let

$$\int_{-\infty}^{\infty} |(Ts_A-s_A T)f(x)|^2\, dx=o(A). \tag{10.08}$$

Then since

$$\lim_{A\to\infty}\frac{1}{2}\int_{-\infty}^{\infty}\left|\frac{Ts_Af(x)}{\sqrt{A}}\right|^2 dx=M_x(|f(x)|^2), \tag{10.09}$$

and

$$\lim_{A\to\infty}\int_{-\infty}^{\infty}\left|\frac{Ts_Af(x)}{\sqrt{A}}-\frac{s_A Tf(x)}{\sqrt{A}}\right|^2 dx=0, \tag{10.10}$$

it follows that

$$M_x(|Tf(x)|^2)=\lim_{A\to\infty}\frac{1}{2}\int_{-\infty}^{\infty}\left|\frac{s_A Tf(x)}{\sqrt{A}}\right|^2 dx=M_x(|f(x)|^2). \tag{10.11}$$

The transformations T form a group. If T_1 and T_2 are of this form,

$$\begin{aligned}
&\int_{-\infty}^{\infty} |(T_1T_2s_A-s_AT_1T_2)f(x)|^2\,dx \\
&=\int_{-\infty}^{\infty} |[(T_1T_2s_A-T_1s_AT_2)+(T_1s_AT_2-s_AT_1T_2)]f(x)|^2\,dx \\
&\leq 2\int_{-\infty}^{\infty} |T_1T_2s_A-T_1s_AT_2)f(x)|^2\,dx+2\int_{-\infty}^{\infty} |(T_1s_A t_2-s_AT_1T_2)f(x)|^2\,dx \\
&=2\int_{-\infty}^{\infty} |(T_2s_A-s_AT_2)f(x)|^2\,dx+2\int_{-\infty}^{\infty} |(T_1s_A-s_AT_1)\,T_2f(x)|^2\,dx.
\end{aligned} \tag{10.12}$$

Furthermore,

$$\int_{-\infty}^{\infty} |(T^{-1}s_A - s_A T^{-1}) f(x)|^2 \, dx = \int_{-\infty}^{\infty} |(s_A - T s_A T^{-1}) f(x)|^2 \, dx$$

$$= \int_{-\infty}^{\infty} |(T s_A - s_A T) T^{-1} f(x)|^2 \, dx. \qquad (10.13)$$

Thus the product of two transformations satisfying (10.08) and the inverse of a transformation satisfying (10.08) likewise satisfy (10.08).

An example of a transformation satisfying (10.08) is

$$T^{\lambda} f(x) = f(x+\lambda), \qquad (10.14)$$

for

$$\int_{-\infty}^{\infty} |(T^{\lambda} s_A - s_A T^{\lambda}) f(x)|^2 \, dx$$

$$\leq \int_{0}^{\lambda} [|f(x-A-\lambda)|^2 + |f(x+A-\lambda)|^2] \, dx = o(A). \qquad (10.15)$$

If T satisfies (10.08) and (10.02) is invariant under it, we shall call it *properly unitary*. Let us consider a one parameter group consisting of all properly unitary transformations U^{λ}, where

$$U^{\lambda+\mu} = U^{\lambda} U^{\mu}. \qquad (10.16)$$

Let $f(x)$ be such that

$$\varphi(t) = M_x [(U^t f(x)) \bar{f}(x)] \qquad (10.17)$$

exists for every t. Clearly, by the Schwarz inequality,

$$\varphi(t) \leq [M_x(|U^t f(x)|^2) M_x(|f(x)|^2)]^{1/2}$$

$$= \left[\varphi(0) \lim_{A \to \infty} \frac{1}{2A} \int_{-\infty}^{\infty} |s_A U^t f(x)|^2 \, dx\right]^{1/2}$$

$$= \left[\varphi(0) \lim_{A \to \infty} \frac{1}{2A} \int_{-\infty}^{\infty} |U^t s_A f(x)|^2 \, dx\right]^{1/2}$$

$$= \varphi(0). \qquad (10.18)$$

It follows that $\varphi(t)$ is a bounded function, and that

$$S(u_1, u_2) = \underset{B\to\infty}{\text{l.i.m.}} \frac{1}{\pi} \int_{-B}^{B} \varphi(t) \frac{\sin(u_2 - u_1)t/2}{t} e^{i\left(\frac{u_1+u_2}{2}\right)t} dt \tag{10.19}$$

exists when $u_2 - u_1$ is given as a quadratically summable function of $u_2 + u_1$. For this one need only apply Plancherel's theorem.

Let us put

$$\varphi_A(t) = \frac{1}{2A} \int_{-\infty}^{\infty} [U^t s_A f(x)]\, \overline{s_A f(x)}\, dx. \tag{10.20}$$

If condition (10.17) holds for every t, we have

$$\varphi(t) = \lim_{A\to\infty} \frac{1}{2A} \int_{-\infty}^{\infty} [s_A U^t f(x)]\, \overline{s_A f(x)}\, dx, \tag{10.21}$$

which we may readily reduce to the form

$$\varphi(t) = \lim_{A\to\infty} \varphi_A(t) \tag{10.22}$$

by means of (10.08). We may easily prove $\varphi_A(t)$ not to exceed $\varphi_A(0)$ and hence to be uniformly bounded in A and t, for

$$\lim_{A\to\infty} \varphi_A(0) = \varphi(0). \tag{10.23}$$

Let us now introduce a new assumption concerning U^x. *Let the transformation W taking $f(x)$ into $U^x f(a)$ (a fixed) preserve* (10.02) *invariant.* Than if

$$\psi(x) = W s_A f(x) \tag{10.24}$$

we have, by the new assumption and (10.20),

$$\varphi_A(t) = \frac{1}{2A} \int_{-\infty}^{\infty} U^i (U^x s_A f(a))\, \overline{U^x s_A f(a)}\, dx$$

$$= \frac{1}{2A} \int_{-\infty}^{\infty} \psi(x+t)\, \overline{\psi}(x)\, dx. \tag{10.25}$$

As we may readily see (cf. (1.29)) $\varphi_A(t)$ is absolutely integrable over $(-\infty, \infty)$, and

$$P_A(u) = \frac{1}{\pi} \int_{-\infty}^{\infty} \varphi_A(t)\, e^{iut}\, dt \tag{10.26}$$

exists, and is real and positive. Indeed, we might have replaced our new assumption by the assumption that

$$\frac{1}{\pi} \int_{-\infty}^{\infty} e^{iut}\, dt \int_{-\infty}^{\infty} [U^t f(x)] f(x)\, dx \tag{10.27}$$

is positive-definite, and exists for every quadratically summable $f(x)$.

Thus it appears that

$$S_A(u_1, u_2) = \frac{2}{\pi} \int_{-\infty}^{\infty} \varphi_A(t) \frac{\sin (u_2 - u_1) t/2}{t} e^{i\left(\frac{u_1+u_2}{2}\right)t}\, dt \tag{10.28}$$

exists, is monotone in u_1 and u_2, and has the property that

$$S_A(u_1, u_2) + S_A(u_2, u_3) = S_A(u_1, u_3). \tag{10.29}$$

To see this, we need only reflect that

$$S_A(u_1, u_2) = \int_{u_1}^{u_2} P_A(u)\, du. \tag{10.30}$$

Now

$$\varphi(t) \frac{\sin (u_2 - u_1) t/2}{t} = \underset{A \to \infty}{\text{l.i.m.}}\ \varphi_A(t) \frac{\sin (u_2 - u_1) t/2}{t}, \tag{10.31}$$

so that

$$S(u_1, u_2) = \underset{A \to \infty}{\text{l.i.m.}}\ S_A(u_1, u_2). \tag{10.32}$$

From this we may readily conclude that

$$S(u_1, u_2) + S(u_2, u_3) = S(u_1, u_3), \tag{10.33}$$

and that $S(u_1, u_2)$ may be so defined as to be monotone in both arguments and increasing in u_2. $S(u_1, u_2)$ is the analogue to $S(u_2) - S(u_1)$ in our earlier sections.

We have

$$\frac{1}{u}\int_0^u S(-u, u)\,du = \frac{1}{\pi u}\int_{-\infty}^{\infty} \varphi(t)\frac{1-\cos ut}{t^2}\,dt. \tag{10.34}$$

Hence

$$S(-\infty, \infty) = \lim_{u\to\infty} \frac{1}{\pi u}\int_{-\infty}^{\infty} \varphi(t)\frac{1-\cos ut}{t^2}\,dt. \tag{10.35}$$

As in (5.47)

$$S(-\infty, \infty) = \lim_{\varepsilon\to 0} \frac{1}{2\varepsilon}\int_{-\varepsilon}^{\varepsilon} \varphi(t)\,dt. \tag{10.36}$$

We thus have arrived at a spectrum theory closely paralleling the theory here developed for trigonometric expansions. Thus for the general case of harmonic analysis, it is the group theory of transformations satisfying (10.08) and (10.24) which is important, rather than the recognized theory of unitary transformations.

Transformations U^t with the properties demanded in this section make their appearance in physics in connection with the Schrödinger equation, which often has its Eigenfunktionen also Eigenfunktionen of an operator analogous to U^t. A more specific instance of $U^t f(x)$ is

$$U^t f(x) = e^{iAt} f(x+t). \tag{10.37}$$

CHAPTER IV.

11. Examples of functions with continuous spectra.

The theory of harmonic analysis which we have so far developed has as one of its purposes the analysis of phenomena such as white light, involving continuous spectra. We have not yet proved our theory to be adequate to this purpose, for we have not yet given a single instance of a continuous spectrum. This lacuna it is the purpose of the present section to fill. To this end, we shall only consider functions $f(t)$ which over every interval $(n, n+1)$, n being an integer, assume one of the two values, 1 and -1. For such a function, the existence of

26—29764. *Acta mathematica.* 55. Imprimé le 15 avril 1930.

$$\varphi(x) = \lim_{T \to \infty} \frac{1}{2T} \int_{-T}^{T} f(t+x) f(t)\, dt$$

for all arguments will follow from its existence for all integral arguments, inasmuch as, if x lies between n and $n+1$,

$$\frac{1}{2T} \int_{-T}^{T} f(t+x) f(t)\, dt = \frac{x-n}{2T} \int_{-T}^{T} f(t+n+1) f(t)\, dt + \frac{n+1-x}{2T} \int_{-T}^{T} f(t+n) f(t)\, dt, \qquad (11.01)$$

so that

$$\varphi(x) = \lim_{T \to \infty} \frac{1}{2T} \int_{-T}^{T} f(t+x) f(t)\, dt = (x-n)\, \varphi(n+1) + (n+1-x)\, \varphi(n). \qquad (11.02)$$

An example of a function of this sort is given below, where the sequence of signs represents the signs of $f(t)$ over the intervals both to the right and the left of the zero point:

+, −;

+, +; +, −; −, +; −, − repeated twice

+, +, +; +, +, −; +, −, +; +, −, −; −, +, +; −, +, −;
−, −, +; −, −, − repeated four times

+, +, +, +; +, +, +, −; +, +, −, +; etc. repeated eight times

. . . . (11.03)

Each repetition of a row is here counted as a separate row. In each row, all the possible arrangements of n symbols which may be either + or − occur arranged in a regular order. Thus in each row, the possible arrangements of p pluses or minuses occur equally often, except for the modification incurred by the possibility that such an arrangement may overlap one of the major divisions indicated by a semicolon in the above table. These semicolons become more and more infrequent as we proceed to later rows in our table, and do not affect the asymptotic distribution of sequence of p signs.

Thus the possible sequences of p signs occur with a probability approaching $1/2^p$ at the end of a comported row. However, the ratio of the number of terms in a row to that in all previous rows approaches zero, so that the effect of stopping at some intermediate point in a row becomes negligible. In other words,

$$\lim \frac{(\text{number of repetitions of a particular sequence of } p \text{ terms in first } n)}{n} = 1/2^p. \qquad (11.04)$$

Hence

$$\lim_{T\to\infty} \frac{1}{2T}\int_{-T}^{T} f(t+m)f(t)\,dt = 0. \qquad [m = \text{an integer} \neq 0] \quad (11.05)$$

Inasmuch as obviously

$$\lim_{T\to\infty} \frac{1}{2T}\int_{-T}^{T} [f(t)]^2\,dt = 1, \qquad (11.06)$$

we see that

$$\varphi(\tau) = \begin{cases} 1 - |\tau|; & [|\tau| \leq 1] \\ 0. & [|\tau| > 1] \end{cases} \qquad (11.07)$$

It follows that

$$S(u) = \frac{1}{2\pi}\int_{-\infty}^{\infty} \varphi(x)\frac{e^{iux}-1}{ix}\,dx = \frac{1}{\pi}\int_{0}^{1}(1-x)\frac{\sin ux}{x}\,dx = \frac{1}{\pi}\int_{0}^{u}\frac{1-\cos v}{v^2}\,dv. \qquad (11.08)$$

Thus the function $f(t)$ has a continuous spectrum with spectral density $\frac{1}{\pi}\frac{1-\cos u}{u^2}$. This fact that the spectrum has a spectral density is an even more restrictive condition than the condition that it should be continuous.

Every monotone function is known to be the sum of three parts: a step function with a denumerable set of steps; a function which is the integral of its derivative; and a continuous function with a derivative almost everywhere zero. This latter type of function has been ignored as a possibility in spectrum analysis. With both line and continuous spectra we are familiar, but the physicists have not considered the possibility of a spectrum in which the total energy of a region tends to zero as the width of the region decreases, but not proportionally in the limit to the width of the region. Nevertheless, functions with a spectrum of this type do exist, as Mr. Kurt Mahler has proved. I am incorporating into this paper an extremely ingenious note of Mr. Mahler, already published in the Journal of Mathematics and Physics of the Massachusetts Institute of Technology, giving an example of this kind.

Let ξ be a simple q-th root of unity, q being any positive integer greater than 1. Let $\bar{\xi}$ be the conjugate complex number, so that

$$\xi\bar{\xi} = 1. \tag{11.09}$$

We define the arithmetical function $\varrho(n)$ by the functional equations

$$\left.\begin{aligned} &\varrho(0) = 1; \\ &\varrho(qn + l) = \xi^l \varrho(n) \text{ for } \begin{bmatrix} l = 0, 1, 2, \ldots, q-1 \\ n = 0, 1, 2, \ldots \end{bmatrix} \end{aligned}\right\} \tag{11.10}$$

We have thus defined $\varrho(n)$ unambiguously for every positive integer n. We may write

$$\varrho(n) = \xi^{q(n)} \tag{11.11}$$

where $q(n)$ is the sum of the digits of n in the q-ary system of notation.

Our problem here is to give an asymptotic evaluation of

$$S_k(n) = \sum_{l=0}^{n-1} \varrho(l)\,\bar{\varrho}(l+k), \tag{11.12}$$

for arbitrary positive integral values of k and large values of n. If $k=0$, we have the obvious formula

$$S_0(n) = n. \tag{11.13}$$

We shall use this as a basis on which to determine

$$S_1(n) = \sum_{l=0}^{n-1} \varrho(l)\,\bar{\varrho}(l+1). \tag{11.14}$$

We may deduce at once from our fundamental equation (11.10) the functional equations of $S_1(n)$, namely

$$\begin{aligned} &S_1(0)=0, \\ &S_1(qn+l)=\bar{\xi}\, S_1(n) + ((q-1)n+l)\,\bar{\xi}. \qquad [l=0, 1, \ldots, q-1] \end{aligned} \tag{11.15}$$

As is obvious, these equations determine $S_1(n)$ unambiguously.

We now see, however, that the series

$$\Sigma_1(n) = \bar{\xi}\left\{[n] - \left[\frac{n}{q}\right]\right\} + \bar{\xi}^2\left\{\left[\frac{n}{q}\right] - \left[\frac{n}{q^2}\right]\right\} + \bar{\xi}^3\left\{\left[\frac{n}{q^2}\right] - \left[\frac{n}{q^3}\right]\right\} + \cdots \quad (11.16)$$

satisfies the same functional equations (11.15) as $S_1(n)$ and hence is identical with $S_1(n)$. We thus have

$$S_1(n) = \bar{\xi}\left\{[n] - \left[\frac{n}{q}\right]\right\} + \bar{\xi}^2\left\{\left[\frac{n}{q}\right] - \left[\frac{n}{q^2}\right]\right\} + \bar{\xi}^3\left\{\left[\frac{n}{q^2}\right] - \left[\frac{n}{q^3}\right]\right\} + \cdots. \quad (11.17)$$

Now let

$$q^r \leq n < q^{r+1}. \quad (11.18)$$

We see that

$$\begin{aligned} S_1(n) &= \bar{\xi}\left\{[n] - \left[\frac{n}{q}\right]\right\} + \bar{\xi}^2\left\{\left[\frac{n}{q}\right] - \left[\frac{n}{q^2}\right]\right\} + \cdots + \bar{\xi}^{r+1}\left\{\left[\frac{n}{q^r}\right] - \left[\frac{n}{q^{r+1}}\right]\right\} \\ &= n\bar{\xi}\left(1 - \frac{1}{q}\right)\left(1 + \frac{\bar{\xi}}{q} + \frac{\bar{\xi}^2}{q^2} + \cdots + \frac{\bar{\xi}^r}{q^r}\right) + O(r) \\ &= n\bar{\xi}\left(1 - \frac{1}{q}\right)\frac{1}{1 - \dfrac{\bar{\xi}}{q}} + O(1) + O(r), \end{aligned} \quad (11.19)$$

or by (11.18)

$$S_1(n) = \frac{q-1}{q\xi - 1} n + O(\log n). \quad (11.20)$$

Formulae (11.13) and (11.20) are only special cases of the corresponding formula for arbitrary k. We obtain this in the following manner.

Since

$$S_k(qn+l) = S_k(qn) + O(1), \qquad [l=0, 1, 2, \ldots, q-1] \quad (11.21)$$

we need only consider $S_k(qn)$. For this we have the formula

$$S_{qK+\lambda}(qn)\,\bar{\xi}^\lambda\,\{(q-\lambda)\,S_K(n) + \lambda\,S_{K+1}(n)\}. \quad (11.22)$$

We define a sequence $\sigma(k)$ by the functional equations

$$\begin{aligned} &\sigma(0) = 1 \\ &\sigma(qK+\lambda) = \bar{\xi}^\lambda\left(\frac{q-\lambda}{q}\sigma(K) + \frac{\lambda}{q}\sigma(K+1)\right). \end{aligned} \quad (11.23)$$

Then it is always true that

$$S_k(n) = \sigma(k)n + O(\log n). \tag{11. 24}$$

To begin with, we have proved this theorem for $k=0$ and $k=1$. Formula (11. 23) shows, however, that we may prove (11. 24) in general by a mathematical induction with respect to k.

$\sigma(k)$ is a very complicated arithmetical function. For small values of its argument ($K=0, 1, \cdots, q-1$, $\lambda=0, 1, \cdots, q-1$) we have

$$\left.\begin{aligned} \sigma(\lambda) &= \bar{\xi}^{\lambda} \frac{(q-\lambda)+(\lambda-1)\bar{\xi}}{q-\bar{\xi}}, \\ \sigma(Kq+\lambda) &= \bar{\xi}^{K+\lambda} \frac{(q+K)(q-\lambda)+((K-1)(q-\lambda)+(q-K-1)\lambda)\bar{\xi}+K\lambda\bar{\xi}^2}{q(q-\bar{\xi})}. \end{aligned}\right\} \tag{11. 25}$$

It is natural to extend our definition of $\sigma(k)$ to negative values of k by the formula

$$\sigma(-k) = \overline{\sigma(k)}. \tag{11. 26}$$

Formula (11. 24) is then true for negative as well as for positive arguments.

It is natural to investigate the functions

$$T_k(n) = \sum_{0}^{n-1} \sigma(l)\,\bar{\sigma}(l+k) \tag{11. 27}$$

which arise from σ in the same fashion as S_k arises from ϱ. We shall confine ourselves to the case

$$q = 2;\ \xi = \bar{\xi} = -1.$$

We have here the equations

$$\left.\begin{aligned} \sigma(k) &= \bar{\sigma}(k); \\ \sigma(2k) &= \sigma(k); \\ \sigma(2k+1) &= -\frac{\sigma(k)+\sigma(k+1)}{2}. \end{aligned}\right\} \tag{11. 28}$$

Hence we have the following formulae:

$$T_{2k}(2n) = \sum_{m=0}^{n-1} (\sigma(2m)\,\sigma(2m+2k) + \sigma(2m+1)\,\sigma(2m+2k+1)) \tag{11.29}$$

$$= \sum_{m=0}^{n-1} \left(\sigma(m)\,\sigma(m+k) + \frac{(\sigma(m)+\sigma(m+1))(\sigma(m+k)+\sigma(m+k+1))}{4}\right) \tag{11.30}$$

or

$$\left| T_{2k}(2n) - \frac{3}{2} T_k(n) - \frac{1}{4} T_{k-1}(n) - \frac{1}{4} T_{k+1}(n) \right| < \text{const.} \tag{11.31}$$

and further

$$T_{2k+1}(2n) = \sum_{m=0}^{n-1} (\sigma(2m)\,\sigma(2m+2k+1) + \sigma(2m+1)\,\sigma(2m+2k+2))$$

$$= -\sum_{m=0}^{n-1} \left(\sigma(m)\frac{\sigma(m+k)+\sigma(m+k+1)}{2} + \frac{\sigma(m)+\sigma(m+1)}{2}\sigma(m+k+1)\right) \tag{11.32}$$

or

$$| T_{2k+1}(2n) + T_k(n) + T_{k+1}(n) | < \text{const.} \tag{11.33}$$

The array

$$\{\ldots, \varrho(n), \ldots, \varrho(1), \varrho(0), \varrho(1), \ldots, \varrho(n), \ldots\} = \{\ldots, a_{-n}, \ldots, a_{-1}, a_0, a_1, \ldots, a_n, \ldots\}$$

defines a function

$$f(t) = \begin{cases} a_n \text{ if } |t-n| \leq 1/8 \\ 0 \text{ if for no } n, |t-n| \leq 1/8 \end{cases} \tag{11.34}$$

and

$$\varphi(\tau) = \lim_{T\to\infty} \frac{1}{2T} \int_{-T}^{T} f(t+\tau)\bar{f}(t)\,dt = \frac{1}{4}\sigma\left(\left[\tau+\frac{1}{4}\right]\right) Q\left(\tau - \left[\tau+\frac{1}{4}\right]\right), \tag{11.35}$$

where

$$Q(x) = \begin{cases} 1-4|x|; & \left[-\frac{1}{4} < x < \frac{1}{4}\right] \\ 0. & \left[\frac{1}{4} \leq x < \frac{3}{4}\right] \end{cases}$$

We have

$$S(u) = \frac{1}{2\pi} \int_{-\infty}^{\infty} \varphi(\tau) \frac{e^{iu\tau}-1}{i\tau}\,d\tau, \tag{11.36}$$

where $S(u)$ is of limited total variation. Then

$$S(u+\varepsilon) - S(u-\varepsilon) = \frac{1}{\pi} \int_{-\infty}^{\infty} \varphi(\tau) \frac{\sin \varepsilon\tau}{\tau} e^{iu\tau}\,d\tau. \tag{11.37}$$

Hence by (6. 15)

$$\lim_{\varepsilon\to 0}\frac{1}{2\varepsilon}\int_{-\infty}^{\infty} e^{-i\nu u}\,|\,S(u+\varepsilon)-S(u-\varepsilon)\,|^2\,du=\lim_{T\to\infty}\frac{1}{2T}\int_{-T}^{T}\varphi(x+\nu)\,\overline{\varphi}(x)\,dx. \qquad (11.\ 38)$$

Hence if the finite or denumerable set of discontinuities of $S(u)$ are at u_1, u_2, ... and have values A_1, A_2, ..., respectively

$$\sum_1^\infty |\,A_k\,|^2=\lim_{\varepsilon\to 0}\frac{1}{2\varepsilon}\int_{-\infty}^{\infty} e^{-i\nu u}\,|\,S(u+\varepsilon)-S(u-\varepsilon)\,|^2\,du=$$

$$=\lim_{T\to\infty}\frac{1}{2T}\int_{-T}^{T}\varphi(x+\nu)\,\overline{\varphi}(x)\,dx \qquad (11.\ 39)$$

and exists for every ν. However,

$$\lim_{T\to\infty}\frac{1}{2T}\int_{-T}^{T}\varphi(x+\nu)\,\overline{\varphi}(x)\,dx=\frac{1}{16}\lim_{n\to\infty}\frac{T_{\left[\nu+\frac{1}{2}\right]}(n)}{n}\,R\left(\nu-\left[\nu+\frac{1}{2}\right]\right), \qquad (11.\ 40)$$

where

$$R(x)=\int_{-\frac{1}{2}}^{\frac{1}{2}} Q(y)\,Q(x+y)\,dy, \qquad (11.\ 41)$$

so that

$$\lim_{n\to\infty}\frac{T_k(n)}{n}$$

exists for every k. If we put

$$\varpi(k)=\lim_{n\to\infty}\frac{T_k(n)}{n}=\lim_{n\to\infty}\frac{1}{n}\sum_{l=0}^{n-1}\sigma(l)\,\sigma(l+k) \qquad (11.\ 42)$$

we may conclude from our equations for T_k that

$$\left.\begin{aligned}\varpi(2k)&=\frac{\varpi(k-1)+6\,\varpi(k)+\varpi(k+1)}{8};\\ \varpi(2k+1)&=-\frac{\varpi(k)+\varpi(k+1)}{2}.\end{aligned}\right\} \qquad (11.\ 43)$$

It follows that if

$$\varpi(0) = 0, \quad \varpi(1) = 0 \tag{11.44}$$

then for every k

$$\varpi(k) = 0. \tag{11.45}$$

We now put $k=0$ in (11.43), remembering that

$$\varpi(-k) = \varpi(k). \tag{11.46}$$

We obtain

$$\varpi(0) = \varpi(1) = 3\,\varpi(1). \tag{11.47}$$

Hence $\varpi(0)=\varpi(1)=0$ and $\varpi(k)$ is identically 0. In other words,

$$T_k(n) = o(n). \tag{11.48}$$

As $\sigma(2k)=\sigma(k)$, $\sigma(1)=-1/3$, we see that we cannot have

$$\lim_{k\to\infty} \sigma(k) = 0, \tag{11.49}$$

and hence we cannot have

$$\lim_{\tau\to\infty} \varphi(\tau) = 0. \tag{11.50}$$

It is thus impossible that

$$S(u) = \int_{-\infty}^{u} S'(v)\,dv + S(-\infty), \tag{11.51}$$

for then we should have by (5.43), (5.46)

$$\lim_{\tau\to\infty} \varphi(\tau) = \lim_{\tau\to\infty} \int_{-\infty}^{\infty} S'(u)\,e^{-iu\tau}\,du = 0. \tag{11.52}$$

On the other hand, as

$$\varpi(0) = 0,$$

we must have

$$\lim_{\varepsilon\to 0} \frac{1}{2\,\varepsilon} \int_{-\infty}^{\infty} |\,S(u+\varepsilon) - S(u-\varepsilon)\,|^2\,du = 0. \tag{11.53}$$

It follows that $S(u)$ is a continuous function which we have already seen not to be the integral of its derivative. This theorem of Mahler thus leads to a new type of spectrum.

27 — 29764. *Acta mathematica.* 55. Imprimé le 15 avril 1930.

12. Spectra depending on an infinite sequence of choices.

In the last section we gave concrete, well defined examples of functions with various continuous types of spectra. The present and the following sections are devoted to the generation of such functions by methods which instead of always yielding functions with continuous spectra, *almost* always yield such functions. The distinction is precisely analogous to that between rational mechanics of the classical kind and statistical mechanics. Theoretically, all the molecules of gas in a vessel might group themselves in a specified half of its volume; practically, such a contingency is unthinkably improbable.

The notion of probability is a new element not occurring in classical mechanics, but essential in statistical mechanics. It applies to a class of contingent situations, and has the essential properties of a measure. So too the idea of »almost always» in harmonic analysis depends on some more or less concealed notion of measure. In the present and the ensuing sections, we shall assimilate this notion of measure to that of Lebesgue, so that »almost always» will translate into »except for a set of Lebesgue measure zero».

Consider a sequence of independent tosses of a coin. By a sequence, we mean a record such as, »heads, tails, heads, heads, tails.» For such a finite sequence, the probability is 2^{-n}, where n is the number of tosses. That is, it is the same as the measure of the set of all the points on $(0, 1)$ with coordinates whose binary expansion begins $.10110$. This mapping immediately suggests a definition of probability for infinite sequences of tosses. The probability of any set of sequences of tosses is defined as the Lebesgue measure of the set of points whose binary representations correspond to sequences of tosses in the set, in such a manner that 1 corresponds to »heads» and 0 to »tails».[5] We can even represent sequences infinite in both directions by binary numbers in such a way as to define the probability of a set of sequences, by having recourse to some definite enumerative rearrangement of such a sequence.

If we have made »probability» a mere translation of »measure», »average» becomes the equivalent of »integral». We are accordingly able to use the entire body of theorems concerning the Lebesgue integral in the service of the calculus of probabilities.

We have not yet, however, correlated with our sequence of throws functions susceptible of a harmonic analysis. To do this, we take a certain zero point on a doubly infinite line to correspond with the zero point of our doubly infinite sequence of tosses, and if the nth toss is a head, we define $f(t)$ to be 1 for $n<t<n+1$; if a tail, to be -1. The question we wish to ask is: what is the probability distribution of spectral functions $S(u)$ for these functions $f(t)$?

We have, taking $f(t)=a_n$ for $n<t<n+1$,

$$\frac{1}{2n}\int_{-n}^{n} f(t+m)f(t)\,dt = \frac{1}{2n}(a_{m-n}\,a_{-n}+a_{m-n+1}\,a_{-n+1}+\cdots+a_{m+n-1}\,a_{n-1}). \quad (12.01)$$

Since the distribution of each a_n between negative and positive values is symmetrical and independent of that of every other,

$$\text{Average } \frac{1}{2n}\int_{-n}^{n} f(t+m)f(t)\,dt = \begin{cases} 0 \text{ if } m \neq 0; \\ 1 \text{ if } m = 0. \end{cases} \quad (12.02)$$

When $m=0$, this average is indeed identically 1. In every other case, when m is an integer,

$$\text{Average } \left\{\frac{1}{2n}\int_{-n}^{n} f(t+m)f(t)\,dt\right\}^2$$

$$= \text{Average } \frac{1}{4n^2}(a_{m-n}\,a_{-n}+\cdots+a_{m+n-1}\,a_{n-1})^2$$

$$= \text{Average } \frac{1}{4n^2}(a^2_{m-n}\,a^2_{-n}+\cdots+a^2_{m+n-1}\,a^2_{n-1})$$

$$= 1/2n, \quad (12.03)$$

since the averages of all the non-square terms vanish. Hence

$$\frac{1}{2n^2}\int_{-n^2}^{n^2} f(t+m)f(t)\,dt > A \quad (12.04)$$

over a set of values of $f(t)$ with total probability

$$\leq \frac{A}{2n^2}. \quad (12.05)$$

Since the sum of these latter quantities forms a convergent series, with remainder after n terms tending to 0 as n increases, we must have

$$\lim_{n\to\infty} \frac{1}{2n^2}\int_{-n^2}^{n^2} f(t+m)f(t)\,dt < A, \tag{12.06}$$

except in a set of cases of arbitrarily small and hence of zero probability. Hence except in a set of cases of zero probability,

$$\lim_{n\to\infty} \frac{1}{2n^2}\int_{-n^2}^{n^2} f(t+m)f(t)\,dt = 0. \tag{12.07}$$

Here the procedure to the limit runs through integral values of n. This can be generalized at once. Let P be bounded, and let

$$n^2 < T < (n+1)^2.$$

Then

$$\frac{1}{2T}\int_{-T}^{T} P\,d\lambda = \frac{n^2}{T}\frac{1}{2n^2}\int_{-n^2}^{n^2} P\,d\lambda + \frac{1}{2T}\left[\int_{n^2}^{T} + \int_{-T}^{-n^2}\right]P\,d\lambda$$

$$= \frac{n^2}{T}\frac{1}{2n^2}\int_{-n^2}^{n^2} P\,d\lambda + \frac{\vartheta}{2T}\left[\int_{-(n+1)^2}^{(n+1)^2} P\,d\lambda - \int_{-n^2}^{n^2} P\,d\lambda\right] \qquad [0 \leq \vartheta \leq 1]$$

$$= \frac{n^2}{T}\frac{1}{2n^2}\int_{-n^2}^{n^2} P\,d\lambda + \vartheta\left[\frac{(n+1)^2}{T}\cdot\frac{1}{2(n+1)^2}\int_{-(n+1)^2}^{(n+1)^2} P\,d\lambda - \frac{n^2}{T}\cdot\frac{1}{2n^2}\int_{-n^2}^{n^2} P\,d\lambda\right]. \tag{12.08}$$

Thus if

$$\lim_{n\to\infty}\frac{1}{2n^2}\int_{-n^2}^{n^2} P\,d\lambda = L,$$

then

$$\lim_{T\to\infty}\frac{1}{2T}\int_{-T}^{T} P\,d\lambda = L. \tag{12.09}$$

We thus see that in case m is an integer other than zero, we almost always have

$$\lim_{T\to\infty} \frac{1}{2T} \int_{-T}^{T} f(t+m) f(t)\, dt = 0. \tag{12. 10}$$

As in (11.07), we may conclude that

$$\varphi(x) = \begin{cases} 0; & (x > 1) \\ 1 - x; & (0 < x < 1) \\ 1 + x; & (-1 < x < 0) \\ 0; & (-1 < x) \end{cases} \tag{12. 11}$$

and that

$$S(u) = \frac{1}{2\pi} \int_{-\infty}^{\infty} \varphi(x) \frac{e^{iux} - 1}{ix}\, dx = \frac{1}{\pi} \int_0^1 (1 - x) \frac{\sin ux}{x}\, dx = \frac{1}{\pi} \int_0^u \frac{1 - \cos v}{v^2}\, dv. \tag{12. 12}$$

These propositions are true, not always, but almost always. Thus a haphazard sequence of positive and negative rectangular impulses almost always has the spectral intensity

$$\frac{1}{\pi} \frac{1 - \cos v}{v^2} \tag{12. 13}$$

which is numerically identical to the square of the Fourier transform of a single rectangular impulse. To see this, we need only reflect that the Fourier transform of such an impulse is

$$\frac{1}{\sqrt{2\pi}} \int_{-\frac{1}{2}}^{\frac{1}{2}} e^{iux}\, dx = \sqrt{\frac{2}{\pi}} \frac{\sin u/2}{u} = \sqrt{\frac{1 - \cos u}{\pi u^2}}. \tag{12. 14}$$

It would not be a difficult task to generalize this remark to impulses of other than rectangular shape. The essential generalization to make this fact of physical interest is, however, to eliminate the equal spacing of the individual impulses, to reduce the sequence of impulses to such an irregularity as is found in the Brownian motion. This is the problem of our next section. The principal difficulty is that the fundamental Lebesgue measure to which we reduce our probabilities is not so obviously at hand. There is no continuous infinity of choices which bears an obvious analogy to that involved in building up a binary fraction. Nevertheless, the distribution involved in the time paths of particles subject to the Brownian motion can be reduced to a Lebesgue measure, certain functions connected with these paths can almost always be analysed harmonically, and their spectra will almost always have a certain fixed distribution of energy if frequency. In other words, the properties of the distributions and functions of this section furnish an excellent working model for those to be expected of the functions discussed in the next section.

13. Spectra and integration in function-space.

From the very beginning, spectrum theory and statistics have joined hands in the theory of white light. The apparent contradiction between the additive character of electromagnetic displacement in the Maxwell theory and the observed additive character of the quadratic light-intensities is on the surface of things irreconcilable. The credit for resolving this antinomy is due to Lord Rayleigh. He considers the resultant of a large number of vibrations of arbitrary phase, and shows that the mean intensity of their sum is actually additive. He says, »It is with this mean intensity only that we are concerned in ordinary photometry. A source of light, such as a candle or even a soda flame, may be regarded as composed of a very large number of luminous centres disposed throughout a very sensible space; and even though it be true that the intensity at a particular point of a screen illuminated by it and at a particular moment of time is a matter of chance, further processes of averaging must be gone through before anything is arrived at of which our senses could ordinarily take cognizance. In the smallest interval of time during which the eye could be impressed, there would be opportunity for any number of rearrangements of phase, due either to motions of the particles or to irregularities in their modes of vibration. And even if we suppose that each luminous centre was fixed, and emitted perfectly regular vibrations, the manner of composition and consequent intensity would vary rapidly from point to point of the screen, and in ordinary cases the mean illumination over the smallest appreciable area would correspond to a thorough averaging of the phase-relationships. In this way the idea of the intensity of a luminous source, independently of any questions of phase, is seen to be justified, and we may properly say that two candles are twice as bright as one.»

Thus Rayleigh's statistics of light is a statistics in which the quantities distributed are amplitudes of sinusoidal vibrations. Such a theory involves a preliminary harmonic analysis, perhaps of a somewhat vague nature, but definite enough to be useful in the hands of a competent physicist. There is an alternative approach to spectrum statistics, which is of a more direct nature. Imagine a resonator — say a sea-shell — struck by a purely chaotic sequence of acoustical impulses. It will yield a response which still has a statistical element in it, but in which the selective properties of the resonator will have accentuated certain frequencies at the expense of others. It seems on the surface of it highly plausible that the outputs of two such resonators will almost always be additive as to intensities rather than merely as to amplitudes.

»Chaos» and »almost always» — there lies an entire statistical theory behind these terms. The simplest phenomenon to which the name »chaos» may be applied with any propriety is that of the Brownian motion. Here a small particle is impelled by atomic collisions in such a way that its future is entirely independent of its past. If we consider the X-coordinate of such a particle, the probability that this should alter a given amount in a given time is independent (1) of the entire past history of the particle; (2) of the instant from which the given interval is measured; (3) whether we are considering changes that increase or changes that decrease it. If we make the assumption that the distribution of the changes of x over a given interval of time is Gaussian, it follows as Einstein has pointed out that the probability that after a time t the change in x should lie between x_1 and $x_1 + dx_1$ is

$$\frac{1}{\sqrt{\pi c t}} e^{-x^2_1/ct} \, dx_1 . \tag{13.01}$$

Here c is a constant which we may and shall reduce to 1 by a proper choice of units. The particular manner in which t enters results from the fact that

$$\frac{1}{\sqrt{\pi(t_1+t_2)}} e^{-\frac{x^2_1}{t_1+t_2}} = \frac{1}{\pi \sqrt{t_1 t_2}} \int_{-\infty}^{\infty} e^{-\frac{x^2}{t_1}} e^{-\frac{(x_1-x)^2}{t_2}} \, dx . \tag{13.02}$$

This fundamental identity is tantamount to the statement that the probability that x should have changed by an amount lying between x_1 and $x_1 + dx_1$ after a time $t_1 + t_2$ is the total compound probability that the change of x over time

t_1 should be anything at all, and that it should then migrate in a subsequent interval of length t_2 to a position between x_1 and x_1+dx_1.

A quantity x whose changes are distributed after the manner just mentioned is said to have them normally or chaotically distributed. Of course, what really is distributed is the function $x(t)$ representing the successive values of x. (There is no essential restriction in supposing $x(0)=0$.) Thus the conception of a purely chaotic distribution really introduces a certain system of measure and consequently of integration into function-space. This gives us the clue to the statistical study of spectra. We determine the response of a resonator in terms of functionals of the incoming chaotic disturbance. We then ask, »What is the distribution of various quantities connected with the spectrum of this response, as determined by integrating these quantities over function-space with respect to the original chaos?» Let it be remarked that the theory of integration appropriate to this problem is that developed by the author in his »Differential-space», and not the earlier theory of Gâteaux, which forms the starting point of most researches in this direction.

Before we can attack these more difficult problems we must establish out theory of integration on a firm basis. To do this, we shall establish a correspondence between the set of all functions and the points on a segment of a line AB of unit length, and shall use this correspondence to define integration over function-space in terms of Lebesgue integration over the segment. Let me say in passing that in my previous attacks on this problem, I have made use of a somewhat generalized form of integration due to P. J. Daniell. This form of integration, at least in all cases yet given, may be mapped into a Lebesgue integration over a one-dimensional manifold by a transformation retaining measure invariant. In as much as the literature contains a much greater wealth of proved theorems concerning the Lebesgue integral than of theorems concerning the Daniell integral — although the latter are not particularly difficult to establish — it has seemed to me more desirable to employ the method of mapping. This method of mapping is an extension to infinitely many dimensions of a method due to Radon.

The method of mapping consists in making certain sets of functions $x(t)$, which we shall call »quasi-intervals», correspond to certain intervals of AB. Our quasi-intervals will be sets of all functions $x(t)$ defined for $0 \leq t \leq 1$ such that

$$\begin{gathered}
x(0)=0;\\
x_{11}\leq x(t_1)\leq x_{12};\\
x_{21}\leq x(t_2)\leq x_{22};\\
x_{31}\leq x(t_3)\leq x_{32};\\
\cdots\cdots\\
x_{n1}\leq x(t_n)\leq x_{n2}.\\
(0\leq t_1\leq t_2\leq t_3\leq\cdots\leq t_n\leq 1).
\end{gathered}\tag{13.03}$$

By our definition of probability, the probability that $x(t)$ should lie in this quasi-interval is

$$\pi^{-n/2}[t_1(t_2-t_1)(t_3-t_2)\cdots(t_n-t_{n-1})]^{-1/2}\int_{x_{11}}^{x_{12}}d\xi_1\cdots\int_{x_{n1}}^{x_{n2}}d\xi_n\cdot$$

$$\cdot\exp\left[-\xi_1^2 t_1^{-1}-\sum_2^n(\xi_k-\xi_{k-1})^2(t_k-t_{k-1})^{-1}\right].\tag{13.04}$$

Clearly, if the class of all functions $x(t)$ be divided into a finite number of quasi-intervals — some of which then must contain infinite values of x_{h1} or x_{h2} — the sum of their probabilities will be 1.

The quasi-intervals with which we shall be more specially concerned are the quasi-intervals $I(n; k_1, k_2, \ldots, k_{2^n})$ for which

$$\left.\begin{aligned}
t_h&=h\,2^{-n}; \qquad (1\leq h\leq 2^n)\\
x_{h1}&=\tan\,(k_h\pi 2^{-n});\\
x_{h2}&=\tan\,((k_h+1)\pi 2^{-n});
\end{aligned}\right\}\tag{13.05}$$

where k_h is some integer between -2^{n-1} and $2^{n-1}-1$, inclusive. For the probability that $x(t)$ should lie in this interval let us write

$$P\{I(n; k_1, \ldots, k_{2^n})\}.$$

Let us notice that $I(n; k_1, \ldots, k_{2^n})$ is made of a finite number of quasi-intervals $I(n; l_1, \ldots, l_{2^{n+1}})$, and that the sum of the probabilities belonging to the latter gives the probability belonging to the former.

Let us now map the four quasi-intervals $I(1; k_1, k_2)$ on the segment AB, assigning to each in order an interval with length equal to its probability. Let us map the quasi-intervals $I(2; k_1, k_2, k_3, k_4)$ into intervals of the segment AB, translating probability into length, and in such a manner that if $I(2; l_1, l_2, l_3, l_4)$ forms a portion of $I(1; k_1, k_2)$, their images stand in the same relation. If we keep this process up indefinitely, we shall have mapped all the quasi-intervals $I(n; k_1, k_2, \ldots, k_{2^n})$ into intervals of AB in such a way that probability is translated into length, and that the end-points of the intervals of AB form an everywhere dense set.

Up to this point, our mapping has mapped quasi-intervals on intervals. We wish to deduce from it a mapping of functions on points. As a lemma for this purpose, we shall show that the functions $x(t)$ for which for any t_1 and t_2 that are terminating binaries,

$$|x(t_1)-x(t_2)| \geq 40\, h\, |t_1-t_2|^{1/4} \tag{13.06}$$

may be enclosed in a denumerable set of quasi-intervals such that the sum of the probabilities of these quasi-intervals is $O(h^{-n})$ for any n.

To show this, let us represent t_1 as the binary fraction

$$0 \cdot a_1\, a_2\, a_3 \cdots a_n \cdots$$

and t_2 as the binary fraction

$$0 \cdot b_1\, b_2\, b_3 \cdots b_n \cdots.$$

Let t_3 be a number whose binary expansion may be made to agree with that of t_1 up to and including a_j and with that of t_2 up to and including b_j. We shall choose t_3 so that j is as large as possible, even though this may necessitate the use of an expression for t_3 ending in $111\cdots$ to agree with the smaller of the quantities t_1 and t_2 and of a terminating expression for t_3 to agree with the larger. The interval from t_1 to t_3 will then be expressible in the form

$$0 \cdot 00 \cdots 0 c_{j+1}\, c_{j+2} \cdots,$$

where there are j consecutive 0's after the final point, and every c is 0 or 1. The interval from t_2 to t_3 may be expressed in a similar manner. In other words, the interval from t_1 to t_2 may be reduced to the sum of a denumerable set of intervals from terminating binaries to adjacent terminating binaries of the same number of figures, such that there are not more than two intervals in the set of magnitude 2^{-j-k} where k is any positive integer, and such that every interval is of one of these sizes.

Now clearly,

$$|t_2-t_1| \geq 2^{-1-j}. \tag{13.07}$$

Hence, if it is for particular values of t_2 and t_1 in question that (13.06) is fulfilled,

$$\begin{aligned} |x(t_1)-x(t_2)| &\geq 40\,h \cdot 2^{-(1+j)/4} \\ &> h \cdot 2^{1-(1+j)/4}/(1-2^{-1/4}) \\ &= 2\,h \sum_{k=j+1}^{\infty} 2^{-l/4}. \end{aligned} \tag{13.08}$$

If we now appeal to our analysis of the interval (t_1, t_2), we see that for *some* interval from $m \cdot 2^{-j-k}$ to $(m+1) \cdot 2^{-j-k}$, where m and k are integers and $0 \leq m < 2^{j+k}$, we shall have

$$\left| x\left(\frac{m}{2^{j+k}}\right) - x\left(\frac{m+1}{2^{j+k}}\right) \right| > h \cdot 2^{-(j+k)/4}. \tag{13.09}$$

Thus if for any pair of values t_1 and t_2 that are terminating binary fractions,

$$|x(t_1)-x(t_2)| \geq 40\,h\,|t_1-t_2|^{1/4}, \tag{13.10}$$

then for some integers m and i $(m < 2^i)$

$$|x(m\,2^{-i})-x((m+1)\,2^{-i})| > h \cdot 2^{-i/4}. \tag{13.11}$$

It merely remains to determine the measure of a denumerable set of our quasi-intervals $I(n; k_1, \ldots, k_{2^n})$ containing all the functions $x(t)$ for which, for some m and i, (13.11) holds.

To begin with, let m and i be fixed. Since our selected quasi-intervals ultimately divide the range of values of $x(m\,2^{-i})$ and of $x((m+1)2^{-i})$ to an arbitrary degree of fineness, there is no trouble in proving that the functions satisfying (13.11) may be enclosed in a finite set of selected quasi-intervals of total probability not exceeding

$$\varepsilon + \frac{2}{\sqrt{\pi 2^{-i}}} \int_{h \cdot 2^{-i/4}}^{\infty} e^{-\frac{x^2}{2^{-i}}}\,dx = \varepsilon + \frac{2}{\sqrt{\pi}} \int_{h \cdot 2^{i/4}}^{\infty} e^{-x^2}\,dx, \tag{13.12}$$

where ε is arbitrarily small. If we sum for m and i, we get as the sum of the probabilities of all our enclosing sets a quantity not exceeding

$$\eta + \sum_{i=1}^{\infty} \frac{2^{i+1}}{\sqrt{\pi}} \int_{h \cdot 2^{i/4}}^{\infty} e^{-x^2}\, dx$$

$$< \sum_{i=1}^{\infty} \frac{2^{i+1}}{\sqrt{\pi}} e^{-h \cdot 2^{i/4}}$$

$$< \sum_{i=1}^{\infty} \frac{2^{i+1-ni} h^{-4n}}{\sqrt{\pi}} \qquad (13.13)$$

for sufficiently large h. As usual η represents an arbitrarily small quantity. Expression (13. 13) clearly can be made to vanish more rapidly than any given negative power of h as h becomes infinite.

Let us now reconsider our mapping. If we leave out the ends of our intervals, which form a denumerable set of measure 0, every point on AB is uniquely characterized by and uniquely characterizes an infinite sequence of intervals, each containing the next, and tending to 0 in length. If we reject a denumerable set of quasi-intervals of arbitrarily small total probability, the remaining quasi-intervals and portions of quasi-intervals all contain functions satisfying the condition of equicontinuity

$$|x(t_1) - x(t_2)| < 40\, h\, |t_1 - t_2|^{1/4}, \qquad (13.14)$$

so that if we modify AB by the removal of a set of points of arbitrarily small outer measure, as well as by the removal of the end-points of our intervals, every point of AB is characterized by a sequence of intervals closing down on it, by the succession of corresponding quasi-intervals, and by the uniquely determined function $x(t)$ common to this sequence of quasi-intervals and satisfying (13. 14). It follows at once that except for a set of points of zero measure, we have determined a unique mapping of the points of AB by functions satisfying (13. 14) for *some* h. Thus any functional of the latter functions determines a function on the line, which may be summable Lebesgue. In the latter case, we shall define the average of the functional as the Lebesgue integral of the corresponding function on AB.

Among the summable functionals are the expressions

$$P(x(t_1), x(t_2), \ldots, x(t_n)),$$

where P stands for a polynomial. This is readily seen to be the case when the expressions $t_1, \ldots, t_n$ are terminating binaries, and the extension to other values follows from the equicontinuity conditions we have already laid down. To see this, let us note that we have already given information enough to prove that the upper average (corresponding to upper integral) of

$$[\max |x(t)|]^n$$

is finite. This functional will, however, simultaneously dominate

$$P(x(t_1), x(t_2), \ldots, x(t_n))$$

in which we suppose P of the nth degree, and the set of approximating functionals

$$\begin{gathered} P(x(t_{11}), x(t_{12}), \ldots, x(t_{1n})); \\ \cdots\cdots\cdots\cdots \\ P(x(t_{m1}), x(t_{m2}), \ldots, x(t_{mn})); \\ \cdots\cdots\cdots\cdots \end{gathered}$$

in which $t_{11}, \ldots, t_{mn}, \ldots$ are terminating binaries, and $\lim_{m\to\infty} t_{mn}=t_n$. That these functionals are actually approximating functionals results from the fact that

$$P(y_1, \ldots, y_n)$$

is continuous, and that $x(t)$ is almost always continuous. Now, there is a theorem to the effect that if a sequence of functions uniformly dominated by a Lebesgue summable function converges for each argument to a limit, and if the Lebesgue integrals of these functions converge to a limit, this limit is the Lebesgue integral of the limit function. This proves our theorem.

In case $t_1 \leq t_2 \leq \cdots \leq t_n$, the average of $P(x(t_1), \ldots, x(t_n))$ is readily seen to be

$$\pi^{-\frac{n}{2}} [t_1(t_2-t_1)\cdots(t_n-t_{n-1})]^{-1/2} \int_{-\infty}^{\infty} d\xi_1 \cdots \int_{-\infty}^{\infty} d\xi_n\, P(\xi_1, \ldots, \xi_n) \cdot \\ \cdot \exp\left[-\xi_1^2 t_1^{-1} - \sum_2^n (\xi_k-\xi_{k-1})^2 (t_k-t_{k-1})^{-1}\right]. \tag{13. 15}$$

In particular, if $t_1 \leq t_2$,

$$\text{Average } (x(t_1)\,x(t_2)) = \frac{1}{\sqrt{\pi t_1(t_2-t_1)}}\int_{-\infty}^{\infty} d\xi_1 \int_{-\infty}^{\infty} \xi_1\,\xi_2\, e^{-\frac{\xi_1^2}{t_1}-\frac{(\xi_2-\xi_1)^2}{t_2-t_1}}\, d\xi_2$$

$$= t_1/2. \tag{13. 16}$$

and if $t_1 \leqq t_2 \leqq t_3 \leq t_4$,

$$\text{Average } (x(t_1)\,x(t_2)\,x(t_3)\,x(t_4)) = \frac{1}{\pi^2 \sqrt{t_1(t_2-t_1)(t_3-t_2)(t_4-t_3)}}\,\cdot$$

$$\cdot \int_{-\infty}^{\infty} d\xi_1 \int_{-\infty}^{\infty} d\xi_2 \int_{-\infty}^{\infty} d\xi_3 \int_{-\infty}^{\infty} d\xi_4\; \xi_1\xi_2\xi_3\xi_4 \exp\left(-\frac{\xi_1^2}{t_1} - \frac{(\xi_2-\xi_1)^2}{t_2-t_1} - \frac{(\xi_3-\xi_2)^2}{t_3-t_2} - \frac{(\xi_4-\xi_3)^2}{t_4-t_3}\right)$$

$$= \frac{t_1\,t_2}{2} + \frac{t_1\,t_3}{4}. \tag{13. 17}$$

The expressions just given are absolutely summable. Accordingly, by the familiar rules for inverting the order of integration, if $\alpha_1(t)$, $\alpha_2(t)$, $\alpha_3(t)$, $\alpha_4(t)$ are of limited total variation over $(0, 1)$,

$$\text{Average} \int_0^1 x(t)\, d\alpha_1(t) \int_0^1 x(t)\, d\alpha_2(t) = \int_0^1 \frac{t}{2}\, d\alpha_1(t) \int_t^1 d\alpha_2(s) + \int_0^1 \frac{t}{2}\, d\alpha_2(t) \int_t^1 d\alpha_1(s)$$

$$= \int_0^1 \frac{t}{2}(\alpha_2(1)-\alpha_2(t))\, d\alpha_1(t) + \int_0^1 \frac{t}{2}(\alpha_1(1)-\alpha_1(t))\, d\alpha_2(t)$$

$$= -\int_0^1 \frac{t}{2}\, d[\alpha_1(1)-\alpha_1(t)][\alpha_2(1)-\alpha_2(t)]$$

$$= \frac{1}{2}\int_0^1 [\alpha_1(1)-\alpha_1(t)][\alpha_2(1)-\alpha_2(t)]\, dt; \tag{13. 18}$$

$$\text{Average} \int_0^1 x(t)\, d\alpha_1(t) \int_0^1 x(t)\, d\alpha_2(t) \int_0^1 x(t)\, d\alpha_3(t) \int_0^1 x(t)\, d\alpha_4(t)$$

$$= \int_0^1 \frac{t}{2}\, d\alpha_1(t) \int_t^1 d\alpha_2(s) \int_s^1 \left(s + \frac{u}{2}\right) d\alpha_3(u) \int_u^1 d\alpha_4(v)$$

$$+\int_0^1 \frac{t}{2}\,d\alpha_1(t)\int_t^1 d\alpha_2(s)\int_s^1\left(s+\frac{u}{2}\right)d\alpha_4(u)\int_u^1 d\alpha_3(v)$$

$+22$ other terms representing different orders of α_1, α_2, α_3, α_4

$$=\frac{1}{2}\int_0^1 \frac{t}{2}\,d\alpha_1(t)\int_t^1 d\alpha_2(s)\int_s^1 \frac{u}{2}\,d\alpha_3(u)\int_u^1 d\alpha_4(v)$$

$$+\frac{1}{2}\int_0^1 \frac{t}{2}\,d\alpha_1(t)\int_t^1 \frac{u}{2}\,d\alpha_3(u)\int_u^1 d\alpha_2(s)\int_s^1 d\alpha_4(v)$$

$$+\frac{1}{2}\int_0^1 \frac{u}{2}\,d\alpha_3(u)\int_u^1 \frac{t}{2}\,d\alpha_1(t)\int_t^1 d\alpha_2(s)\int_s^1 d\alpha_4(v)$$

$$+\frac{1}{2}\int_0^1 \frac{u}{2}\,d\alpha_3(u)\int_u^1 d\alpha_4(v)\int_v^1 \frac{t}{2}\,d\alpha_1(t)\int_t^1 d\alpha_2(s)$$

$$+\frac{1}{2}\int_0^1 \frac{u}{2}\,d\alpha_3(u)\int_u^1 \frac{t}{2}\,d\alpha_1(t)\int_t^1 d\alpha_4(v)\int_v^1 d\alpha_2(s)$$

$$+\frac{1}{2}\int_0^1 \frac{t}{2}\,d\alpha_1(t)\int_t^1 \frac{u}{2}\,d\alpha_3(u)\int_u^1 d\alpha_4(v)\int_v^1 d\alpha_2(s)$$

$+$ all other terms representing different orders of α_1, α_2, α_3, α_4

$$=\frac{1}{2}\int_0^1 \frac{t}{2}\,d\alpha_1(t)\int_t^1 d\alpha_2(s)\int_0^1 \frac{u}{2}\,d\alpha_3(u)\int_u^1 d\alpha_4(v)+\text{etc.}$$

$$=\frac{1}{4}\int_0^1 [\alpha_1(1)-\alpha_1(t)][\alpha_2(1)-\alpha_2(t)]\,dt\int_0^1 [\alpha_3(1)-\alpha_3(s)][\alpha_4(1)-\alpha_4(s)]\,ds$$

$$+\frac{1}{4}\int_0^1 [\alpha_1(1)-\alpha_1(t)][\alpha_3(1)-\alpha_3(t)]\,dt\int_0^1 [\alpha_2(1)-\alpha_2(s)][\alpha_4(1)-\alpha_4(s)]\,ds$$

$$+\frac{1}{4}\int_0^1 [\alpha_1(1)-\alpha_1(t)][\alpha_4(1)-\alpha_4(t)]\,dt\int_0^1 [\alpha_2(1)-\alpha_2(s)][\alpha_3(1)-\alpha_3(s)]\,ds. \quad (13.19)$$

The point of this last argument is that

$$\frac{t}{2}\left(s+\frac{u}{2}\right)$$

may be written

$$\frac{t}{2}\frac{s}{2}+\frac{s}{2}\frac{t}{2}+\frac{t}{2}\frac{u}{2}$$

and that we may then take advantage of the existence in our expression of all permutations of $\alpha_1, \alpha_2, \alpha_3, \alpha_4$ to relabel our variables s, t, u, v so as to add the terms of our expression together again in a more symmetrical way, and represent it as a sum of three products of integrals such as we have already evaluated.

Up to the present, we have been considering probabilities depending on a basis function $x(t)$ defined over $(0, 1)$. For the purposes of harmonic analysis, we wish to replace this by a basis function defined over $(-\infty, \infty)$. We may do this as follows: Let

$$\xi(\tau)=\sqrt{\pi}\left[\int_{\frac{1}{2}}^{\frac{1}{\pi}\cot^{-1}(-\tau)} x(t)\,d\csc\pi t-x\left(\frac{1}{\pi}\cot^{-1}(-\tau)\right)\sqrt{1+\tau^2}+x(1/2)\right]. \qquad (13.20)$$

As $x(t)$ is almost always bounded, this will almost always exist. Then

$$\text{Average } (\xi(\beta)-\xi(\alpha))^2=\frac{\pi}{2}\int_{\frac{1}{\pi}\cot^{-1}(-\alpha)}^{\frac{1}{\pi}\cot^{-1}(-\beta)}\csc^2\pi t\,dt=\frac{\beta-\alpha}{2}, \qquad (13.21)$$

in the case that $\beta>\alpha$. This is merely a particular case of (13.18). In the case that (α, β) and (γ, δ) do not overlap, a similar argument will show that

$$\text{Average } (\xi(\beta)-\xi(\alpha))(\xi(\gamma)-\xi(\delta))=0. \qquad (13.22)$$

Thus $\xi(\tau)$ has essentially the same distribution properties as $x(t)$, but over $(-\infty, \infty)$ instead of $(0, 1)$.

We might, of course, have defined our distribution of $\xi(\tau)$ originally, without any recourse to that of $x(t)$. In any case, we should have had to make use of the fact that this distribution has certain equicontinuity properties, and these are somewhat easier to develop over a finite than over an infinite interval. The function $\xi(\tau)$ represents the result of a haphazard sequence of impulses, uniformly distributed in time, extending from $-\infty$ to ∞, in such a way that their zero value is taken to be at $\tau=0$. It is consequently immediately available for a harmonic analysis such as we discuss in this paper, while $x(t)$ is itself immediately adapted only for a Fourier series analysis.

Now let $\vartheta(\tau)$ represent the characteristic response in time of some resonator, the so-called indicial admittance. It may be real or complex, but we shall assume $\vartheta(\tau)\sqrt{1+\tau^2}$ to be of limited total variation over $(-\infty, \infty)$ and ϑ to be quadratically summable. As an immediate consequence of these assumptions,

$$\vartheta(-\infty)=\vartheta(\infty)=0. \tag{13.23}$$

We have

$$\int_{-\infty}^{\infty}\xi(\tau)\,d\vartheta(\tau)=\sqrt{\pi}\int_{-\infty}^{\infty}d\vartheta(\tau)\left[\int_{\frac{1}{2}}^{\frac{1}{\pi}\cot^{-1}(-\tau)}x(t)\,d\csc t-x\left(\frac{1}{\pi}\cot^{-1}(-\tau)\right)\sqrt{1+\tau^2}+x(1/2)\right]$$

$$=\sqrt{\pi}\int_0^1 x(t)[\vartheta(-\cot\pi t)\,d\csc\pi t+\csc\pi t\,d\vartheta(-\cot\pi t)]$$

$$=\sqrt{\pi}\int_0^1 x(t)\,d[\vartheta(-\cot\pi t)\csc\pi t]. \tag{13.24}$$

Hence if $\vartheta_1(\tau)$, $\vartheta_2(\tau)$, $\vartheta_3(\tau)$, and $\vartheta_4(\tau)$ satisfy the conditions we have laid down for $\vartheta(\tau)$,

$$\text{Average}\int_{-\infty}^{\infty}\xi(\tau)\,d\vartheta_1(\tau)\int_{-\infty}^{\infty}\xi(\tau)\,d\vartheta_2(\tau)=\frac{1}{2}\int_0^1\pi\csc^2\pi t\,\vartheta_1(-\cot\pi t)\vartheta_2(-\cot\pi t)\,dt$$

$$=\frac{1}{2}\int_{-\infty}^{\infty}\vartheta_1(\tau)\vartheta_2(\tau)\,d\tau; \tag{13.25}$$

$$\text{Average} \int_{-\infty}^{\infty} \xi(\tau)\, d\vartheta_1(\tau) \int_{-\infty}^{\infty} \xi(\tau)\, d\vartheta_2(\tau) \int_{-\infty}^{\infty} \xi(\tau)\, d\vartheta_3(\tau) \int_{-\infty}^{\infty} \xi(\tau)\, d\vartheta_4(\tau)$$

$$= \frac{1}{4} \int_{-\infty}^{\infty} \vartheta_1(\tau)\vartheta_2(\tau)\, d\tau \int_{-\infty}^{\infty} \vartheta_3(\tau)\vartheta_4(\tau)\, d\tau$$

$$+ \frac{1}{4} \int_{-\infty}^{\infty} \vartheta_1(\tau)\vartheta_3(\tau)\, d\tau \int_{-\infty}^{\infty} \vartheta_2(\tau)\vartheta_4(\tau)\, d\tau$$

$$+ \frac{1}{4} \int_{-\infty}^{\infty} \vartheta_1(\tau)\vartheta_4(\tau)\, d\tau \int_{-\infty}^{\infty} \vartheta_2(\tau)\vartheta_3(\tau)\, d\tau. \qquad (13.26)$$

We thus have succeeded in generalizing our theorems concerning the averages of products of linear functionals to the case where the basis function has an infinite range.

We wish to apply these results to the harmonic analysis of an expression $\int_{-\infty}^{\infty} \xi(\tau)\, d\vartheta(\tau+\lambda)$. To do this, we must evaluate the following averages:

$$\text{Average} \frac{1}{2T} \int_{-T}^{T} d\lambda \int_{-\infty}^{\infty} \xi(\tau)\, d\vartheta(\tau+\lambda) \int_{-\infty}^{\infty} \xi(\sigma)\, d\bar{\vartheta}(\sigma+\lambda)$$

$$= \frac{1}{2T} \int_{-T}^{T} d\lambda \; \text{Average} \left[\int_{-\infty}^{\infty} \xi(\tau)\, d\vartheta(\tau+\lambda) \int_{-\infty}^{\infty} \xi(\sigma)\, d\bar{\vartheta}(\sigma+\lambda) \right]$$

$$= \frac{1}{4T} \int_{-T}^{T} d\lambda \int_{-\infty}^{\infty} |\vartheta(\tau+\lambda)|^2\, d\tau$$

$$= \frac{1}{2} \int_{-\infty}^{\infty} |\vartheta(\tau)|^2\, d\tau. \qquad (13.27)$$

(Here as in what follows, the inversion of the operators $\frac{1}{2T}\int_{-T}^{T} d\tau$ and Average is permissable, since the integral to which our mapping process leads us is absolutely convergent.)

$$\text{Average}\left[\frac{1}{2T}\int_{-T}^{T}d\lambda\int_{-\infty}^{\infty}\xi(\tau)\,d\vartheta(\tau+\lambda)\int_{-\infty}^{\infty}\xi(\sigma)\,d\bar{\vartheta}(\sigma+\lambda)-\frac{1}{2}\int_{-\infty}^{\infty}|\vartheta(\tau)|^2\,d\tau\right]^2$$

$$=\text{Average}\,\frac{1}{4T^2}\int_{-T}^{T}d\lambda\int_{-T}^{T}d\mu\int_{-\infty}^{\infty}\xi(\tau)\,d\vartheta(\tau+\lambda)\int_{-\infty}^{\infty}\xi(\sigma)\,d\bar{\vartheta}(\sigma+\lambda)\int_{-\infty}^{\infty}\xi(\alpha)\,d\vartheta(\alpha+\mu)\,\cdot$$

$$\int_{-\infty}^{\infty}\xi(\beta)\,d\bar{\vartheta}(\beta+\mu)-\frac{1}{4}\left[\int_{-\infty}^{\infty}|\vartheta(\tau)|^2\,d\tau\right]^2$$

$$=\frac{1}{4T^2}\int_{-T}^{T}d\lambda\int_{-T}^{T}d\mu\,\frac{1}{4}\left\{\left|\int_{-\infty}^{\infty}\vartheta(\tau+\lambda)\bar{\vartheta}(\tau+\mu)\,d\tau\right|^2+\left|\int_{-\infty}^{\infty}\vartheta(\tau+\lambda)\vartheta(\tau+\mu)\,d\tau\right|^2\right\}$$

$$\leq\frac{1}{32T^2}\int_{-2T}^{2T}d\lambda\int_{-2T}^{2T}du\left\{\left|\int_{-\infty}^{\infty}\vartheta(\tau)\bar{\vartheta}(\tau+u)\,d\tau\right|^2+\left|\int_{-\infty}^{\infty}\vartheta(\tau)\vartheta(\tau+u)\,d\tau\right|^2\right\}$$

$$\leq\frac{1}{8T}\int_{-\infty}^{\infty}du\left\{\left|\int_{-\infty}^{\infty}\vartheta(\tau)\bar{\vartheta}(\tau+u)\,d\tau\right|^2+\left|\int_{-\infty}^{\infty}\vartheta(\tau)\vartheta(\tau+u)\,d\tau\right|^2\right\}. \tag{13.28}$$

The function in the bracket in the last expression will be summable since $\vartheta(\tau)$ is, as we see from (1.28).

It follows that for any positive number A,

$$\left|\frac{1}{2T}\int_{-T}^{T}d\lambda\left|\int_{-\infty}^{\infty}\xi(\tau)\,d\vartheta(\tau+\lambda)\right|^2-\frac{1}{2}\int_{-\infty}^{\infty}|\vartheta(\tau)|^2\,d\tau\right|<A \tag{13.29}$$

except for a set of values of $x(t)$ not exceeding

$$\frac{1}{8TA^2}\int_{-\infty}^{\infty}du\left\{\left|\int_{-\infty}^{\infty}\vartheta(\tau)\bar{\vartheta}(\tau+u)\,d\tau\right|^2+\left|\int_{-\infty}^{\infty}\vartheta(\tau)\vartheta(\tau+u)\,d\tau\right|^2\right\} \tag{13.30}$$

in outer measure. Let T now assume the successive values 1, 4, 9, ... Then the probability that (13.29) fails to be satisfied for some T from $1/n^2$ on does not exceed the remainder of the convergent series

$$\frac{1}{8A^2}\int_{-\infty}^{\infty} du \left\{\left|\int_{-\infty}^{\infty} \vartheta(\tau)\bar{\vartheta}(\tau+u)\,d\tau\right|^2 + \left|\int_{-\infty}^{\infty} \vartheta(\tau)\vartheta(\tau+u)\,d\tau\right|^2\right\}\left[1+\frac{1}{4}+\frac{1}{9}+\cdots\right]. \quad (13.31)$$

Inasmuch as this remainder is arbitrarily small, we almost always have

$$\overline{\lim_{n\to\infty}}\left|\frac{1}{2n^2}\int_{-n^2}^{n^2} d\lambda \left|\int_{-\infty}^{\infty} \xi(\tau)\,d\vartheta(\tau+\lambda)\right|^2 - \frac{1}{2}\int_{-\infty}^{\infty} |\vartheta(\tau)|^2\,d\tau\right| \leq A. \quad (13.32)$$

Since, however, A is an arbitrary positive quantity,

$$\lim_{n\to\infty}\left|\frac{1}{2n^2}\int_{-n^2}^{n^2} d\lambda \left|\int_{-\infty}^{\infty} \xi(\tau)\,d\vartheta(\tau+\lambda)\right|^2 - \frac{1}{2}\int_{-\infty}^{\infty} |\vartheta(\tau)|^2\,d\tau\right| = 0. \quad (13.33)$$

As in the preceding section, we may conclude that

$$\lim_{T\to\infty}\frac{1}{2T}\int_{-T}^{T} d\lambda \left|\int_{-\infty}^{\infty} \xi(\tau)\,d\vartheta(\tau+\lambda)\right|^2 = \frac{1}{2}\int_{-\infty}^{\infty} |\vartheta(\tau)|^2\,d\tau, \quad (13.34)$$

except in a set of cases of zero probability.

Let us now consider

$$\int_{-A}^{A} d\lambda \int_{-\infty}^{\infty} \xi(\tau)\,d\vartheta(\tau+\lambda) = \int_{-\infty}^{\infty} \xi(\tau)\,d\int_{-A}^{A} \vartheta(\tau+\lambda)\,d\lambda, \quad (13.35)$$

for rational values of A, $\vartheta(\tau)$ being subject to the conditions already laid down. Let us put

$$\int_{-\infty}^{\infty} \xi(\tau)\,d\vartheta(\tau+\lambda) = f(\lambda). \quad (13.36)$$

We have almost always, for any denumerable set of values of A, as for example, for all rational values of A,

$$\lim_{T\to\infty} \frac{1}{2T}\int_{-T}^{T} d\lambda \left|\int_{-A}^{A} f(\lambda+\mu)\,d\mu\right|^2 = \frac{1}{2}\int_{-\infty}^{\infty} d\tau \left|\int_{-A}^{A} \vartheta(\tau+\mu)\,d\mu\right|^2. \tag{13.37}$$

This results from the fact that the sum of a denumerable set of null sets is a null set. As before, let us put

$$s(u) = \frac{1}{2\pi}\int_{-1}^{1} f(x)\,\frac{e^{ixu}-1}{ix}\,dx + \frac{1}{2\pi}\,\underset{M\to\infty}{\mathrm{l.i.m.}}\left[\int_{1}^{M} + \int_{-M}^{-1}\right] f(x)\,\frac{e^{ixu}}{ix}\,dx. \tag{13.38}$$

It then follows from (13. 34) and (13. 37), with the help of (6. 23), that we shall almost always have

$$\lim_{\varepsilon\to 0} \frac{1}{2\varepsilon}\int_{-\infty}^{\infty} |s(u+\varepsilon)-s(u-\varepsilon)|^2\,du = \frac{1}{2}\int_{-\infty}^{\infty} |\vartheta(\tau)|^2\,d\tau, \tag{13.39}$$

and (for all rational A)

$$\lim_{\varepsilon\to 0} \frac{1}{2\varepsilon}\int_{-\infty}^{\infty} \frac{4\sin^2 Au}{u^2}\,|s(u+\varepsilon)-s(u-\varepsilon)|^2\,du = \frac{1}{2}\int_{-\infty}^{\infty} d\tau \left|\int_{-A}^{A} \vartheta(\tau+\mu)\,d\mu\right|^2. \tag{13.40}$$

Now let us put

$$\vartheta(\tau) = \frac{1}{\sqrt{2\pi}}\,\underset{M\to\infty}{\mathrm{l.i.m.}}\int_{-M}^{M} \psi(u)\,e^{iu\tau}\,du. \tag{13.41}$$

Then

$$\int_{-A}^{A} \vartheta(\tau+\mu)\,d\mu = \sqrt{\frac{2}{\pi}}\int_{-\infty}^{\infty} \psi(u)\,\frac{\sin Au}{u}\,e^{iu\tau}\,du. \tag{13.42}$$

Thus if $\vartheta(\tau)\sqrt{1+\tau^2}$ is of limited total variation and $\vartheta(\tau)$ is quadratically summable, we almost always have for all rational A

$$\int_{-\infty}^{\infty} d\tau \left|\int_{-A}^{A} \vartheta(\tau+\mu)\,d\mu\right|^2 = \int_{-\infty}^{\infty} |\psi(u)|^2\,\frac{4\sin^2 Au}{u^2}\,du. \tag{13.43}$$

In other words, we almost always have for all rational A,

$$\lim_{\varepsilon\to 0}\int_{-\infty}^{\infty}\frac{\sin^2 Au}{u^2}\left\{\frac{1}{\varepsilon}|s(u+\varepsilon)-s(u-\varepsilon)|^2-|\psi(u)|^2\right\}du=0, \qquad (13.44)$$

and

$$\lim_{\varepsilon\to 0}\int_{-\infty}^{\infty}\left\{\frac{1}{\varepsilon}|s(u+\varepsilon)-s(u-\varepsilon)|^2-|\psi(u)|^2\right\}du=0. \qquad (13.45)$$

Thus we almost always have

$$\lim_{\varepsilon\to 0}\int_{-\infty}^{\infty}P(u)\left\{\frac{1}{\varepsilon}|s(u+\varepsilon)-s(u-\varepsilon)|^2-|\psi(u)|^2\right\}du=0, \qquad (13.46)$$

in case

$$P(u)=\sum_{1}^{n}\frac{A_k\sin^2 N_k(u)}{u^2} \qquad (13.47)$$

it follows from (13.45) that we may even replace $P(u)$ by

$$Q(u)=\underset{n\to\infty}{\text{uniform limit}}\ P_n(u) \qquad (13.48)$$

where $P_n(u)$ is of the form given above for $P(u)$. Thus by the Weierstrass theorem, $Q(u)$ may be the quotient by u^2 of any continuous periodic function with any period. Since we can approximate by such a function Q to any continuous function vanishing at $\pm\infty$, our sole condition on Q may be replaced by

$$Q(u)=o(1) \quad \text{at} \quad u=\pm\infty. \qquad (13.49)$$

Even this does not represent the utmost extension of our theorem. It follows at once by subtracting from 1 a Q vanishing outside of a finite range that

$$\lim_{N\to\infty}\ \overline{\lim_{\varepsilon\to 0}}\left[\int_{N}^{\infty}+\int_{-\infty}^{-N}\right]\left\{\frac{1}{\varepsilon}|s(u+\varepsilon)-s(u-\varepsilon)|^2-|\psi(u)|^2\right\}du=0. \qquad (13.50)$$

Thus a bounded modification of Q for large arguments produces a decreasing effect as the range of modification recedes to infinity, and we have as our sole condition to be imposed on the continuous function Q that

$$Q(u)=O(1) \quad \text{at} \quad u=\pm\infty. \qquad (13.51)$$

A case of peculiar importance is where $e^{-ivu}=Q(u)$. Here

$$\lim_{\varepsilon \to 0} \frac{1}{2\varepsilon} \int_{-\infty}^{\infty} e^{-i\nu u} |s(u+\varepsilon) - s(u-\varepsilon)|^2 du = \frac{1}{2} \int_{-\infty}^{\infty} e^{-i\nu u} |\psi(u)|^2 du. \quad (13.52)$$

This exists for *every* ν for almost all $x(t)$. Thus by (6.15),

$$\varphi(\nu) = \frac{1}{2} \int_{-\infty}^{\infty} e^{-i\nu u} |\psi(u)|^2 du, \quad (13.53)$$

and $\varphi(\nu)$ exists for *every* ν for almost every $x(t)$. As in section 3, let us put

$$S(u) = \frac{1}{2\pi} \int_{-\infty}^{\infty} \varphi(x) \frac{e^{iux} - 1}{ix} dx, \quad (13.54)$$

so that we get by (5.42), (5.45)

$$\varphi(x) = \int_{-\infty}^{\infty} e^{-ixu} dS(u). \quad (13.55)$$

We have already shown φ and S to exist on the assumption that $\varphi(\nu)$ exists for every ν. We have, by (13.53) and (13.55), for almost all $x(t)$

$$\int_{-\infty}^{\infty} e^{-i\nu u} d \left[S(u) - \frac{1}{2} \int_{-\infty}^{u} |\psi(v)|^2 dv \right] \equiv 0. \quad (13.56)$$

By processes now familiar (cf. (13.46)), we can replace $e^{-i\nu u}$ by functions Q which are merely continuous and bounded. We here make use of the absolute convergence of (13.56). Hence the average of $S(u) - \frac{1}{2} \int_{-\infty}^{u} |\psi(v)|^2 dv$ vanishes over every interval, and

$$S(u) - \frac{1}{2} \int_{-\infty}^{u} |\psi(v)|^2 dv \equiv 0; \quad (13.57)$$

and consequently

$$S'(u) = \frac{1}{2} |\psi(u)|^2, \quad (13.58)$$

except for a set of values of a zero measure. Inasmuch as $S'(u)$ is the spectral density of $f(x)$, we see that as a consequence of our assumptions that $\vartheta(\tau)$ is quadratically summable and of limited total variation when multiplied by $\sqrt{1+\tau^2}$ *the spectral density of* $\int_{-\infty}^{\infty} \xi(\tau) d\vartheta(\tau+\lambda)$ *is half the square of the modulus of the Fourier transform of* ϑ. Another way of phrasing this fact is: *if a linear resonator is stimulated by a uniformly haphazard sequence of impulses, each frequency responds with an amplitude proportional to that which it would have if stimulated by an impulse of that frequency and of unit energy.* An even simpler statement is: *the energy of a haphazard sequence of impulses is uniformly distributed in frequency.* This law of distribution bears a curious analogy to that predicted for white light by the incorrect Boltzmann law of radiation. The physical conditions which lead to this law of distribution of energy in frequency are that the sequence of impulses in question should be distributed over every interval of time in a Gaussian manner, that their past should not influence their future, that very many should occur over the smallest period of time to be investigated, and that the modulus of the Gaussian distribution of these impulses for a given time interval should depend only on the length of this interval. These conditions are approximately realized in the case of the Schroteffekt, where an electrical resonating circuit is set in vibration by the irregularities in the stream of electrons across a vacuum tube. It might also be realized in the ease of an acoustical system set in oscillation by such a noise as that of a sand blast. Theoretically this equipartition of energy might be used in the absolute calibration of acoustical instruments.

Just as the average of an expression depending on a single function $x(t)$ may be reduced a Lebesgue single integral, so a similar average depending on two independent functions $x(t)$ and $y(t)$ may be reduced to a Lebesgue double integral. On the assumption that $\vartheta_1(\tau)$ and $\vartheta_2(\tau)$ satisfy the conditions we have already laid down for $\vartheta(\tau)$ and that

$$\left.\begin{aligned} \xi(\tau) &= \sqrt{\pi} \int_{1/2}^{\frac{1}{\pi}\cot^{-1}(-\tau)} x(t)\, d \csc \pi t - x\left(\frac{1}{\pi}\cot^{-1}(-\tau)\right) + x(1/2); \\ \eta(\tau) &= \sqrt{\pi} \int_{1/2}^{\frac{1}{\pi}\cot^{-1}(-\tau)} y(t)\, d \csc \pi t - y\left(\frac{1}{\pi}\cot^{-1}(-\tau)\right) + y(1/2); \end{aligned}\right\} \quad (13.59)$$

$$\vartheta_{1,2}(\tau) = \frac{1}{\sqrt{2\pi}} \underset{M\to\infty}{\text{l.i.m.}} \int_{-M}^{M} \psi_{1,2}(u) e^{iu\tau} du \tag{13.60}$$

it is easy to prove by methods substantially identical with those already employed that

$$\lim_{T\to\infty} \frac{1}{2T} \int_{-T}^{T} \left| \int_{-\infty}^{\infty} \xi(\tau) d\vartheta_1(\tau+\lambda) + \int_{-\infty}^{\infty} \eta(\tau) d\vartheta_2(\tau+\lambda) \right|^2 d\lambda \tag{13.61}$$

almost always has a certain definite value. Inasmuch as a normal distribution for $x(t)$ implies the same for $-x(t)$,

$$\lim_{T\to\infty} \frac{1}{2T} \int_{-T}^{T} \left| -\int_{-\infty}^{\infty} \xi(\tau) d\vartheta_1(\tau+\lambda) + \int_{-\infty}^{\infty} \eta(\tau) d\vartheta_2(\tau+\lambda) \right|^2 d\lambda \tag{13.62}$$

almost always has the same value. Subtracting, we almost always have

$$\lim_{T\to\infty} \frac{1}{2T} \int_{-T}^{T} \Re \left[\int_{-\infty}^{\infty} \xi(\tau) d\vartheta_1(\tau+\lambda) \int_{-\infty}^{\infty} \eta(\tau) d\bar{\vartheta}_2(\tau+\lambda) \right] d\lambda = 0. \tag{13.63}$$

If we work in a similar manner with

$$\lim_{T\to\infty} \frac{1}{2T} \int_{-T}^{T} \left| \int_{-\infty}^{\infty} \xi(\tau) d\vartheta_1(\tau+\lambda) \pm i \int_{-\infty}^{\infty} \eta(\tau) d\vartheta_2(\tau+\lambda) \right|^2 d\lambda, \tag{13.64}$$

we see that almost always

$$\lim_{T\to\infty} \frac{1}{2T} \int_{-T}^{T} \Im \left[\int_{-\infty}^{\infty} \xi(\tau) d\vartheta_1(\tau+\lambda) \int_{-\infty}^{\infty} \eta(\tau) d\bar{\vartheta}_2(\tau+\lambda) \right] d\lambda = 0. \tag{13.65}$$

Hence almost always

$$\lim_{T\to\infty} \frac{1}{2T} \int_{-T}^{T} \left[\int_{-\infty}^{\infty} \xi(\tau) d\vartheta_1(\tau+\lambda) \int_{-\infty}^{\infty} \eta(\tau) d\bar{\vartheta}_2(\tau+\lambda) \right] d\lambda = 0 \tag{13.66}$$

and the coherency matrix of $\int_{-\infty}^{\infty} \xi(\tau)\, d\vartheta_1(\tau+\lambda)$ and $\int_{-\infty}^{\infty} \eta(\tau)\, d\vartheta_2(\tau+\lambda)$ is almost always

$$\left\| \begin{matrix} \frac{1}{2}|\psi_1(u)|^2 & 0 \\ 0 & \frac{1}{2}|\psi_2(u)|^2 \end{matrix} \right\|. \tag{13.67}$$

As a direct consequence, if the motion of a particle is independently haphazard in two directions at right angles, and if this motion influences a resonator with the same characteristics in the two directions, the coherency matrix of the motion of the resonator is unpolarized.

In the opinion of the author, the chief importance of this section is in showing in a systematic manner how the Lebesgue integral may be adapted to the needs of statistical mechanics. It is no new observation that sets of zero measure and sets of phenomena, not necessarily impossible, of probability zero, are in essence the same sort of thing. It is not, however, a particularly easy matter to translate any specific problem in statistical mechanics into its precise counterpart in the theory of integration. The author feels confident that methods closely resembling those here developed are destined to play a part in the statistical mechanics of the future, in such regions as those now being invaded by the theory of quanta.

CHAPTER V.

14. The spectrum of an almost periodic function.

The last paragraph was exclusively devoted to functions with continuous spectra; we now come to the most important known class of functions with spectra that are discrete. This is the class of almost periodic functions, the discovery of which is due to Harald Bohr. Let $f(x)$ be a continuous function, not necessarily real, defined for all real values of x between $-\infty$ and ∞. If ε is any positive quantity, Bohr defines τ_ε to be a translation number of $f(x)$ belonging to ε, in case for every real x,

$$|f(x+\tau_\varepsilon)-f(x)| \leq \varepsilon. \tag{14.01}$$

In case, whenever, ε is given, a quantity L_ε can be assigned, such that no interval $(a, a+L_\varepsilon)$ is free of translation numbers τ_ε belonging to ε, $f(x)$ is said to be *almost periodic*. Bohr's most fundamental theorem is: *the necessary and sufficient condition for a function $f(x)$ to be almost periodic is that for any positive quantity ε, there exist a finite set of complex numbers $A_1, A_2, \ldots, A_n$ and a set of real numbers $\Lambda_1, \Lambda_2, \ldots, \Lambda_n$, such that for all x*

$$\left| f(x) - \sum_{1}^{n} A_k e^{i\Lambda_k x} \right| < \varepsilon. \tag{14.02}$$

The next few sections of this paper are devoted to the proof of this theorem. In this proof we shall avail ourselves of the following theorems of Bohr concerning almost periodic functions, which are susceptible of a completely elementary proof:

Any finite set of almost periodic functions is *simultaneously* almost periodic, in the sense that for any ε, L_ε may be assigned for the whole set at once, in such a manner that in any interval $(a, a+L_\varepsilon)$, there is at least one translation number τ_ε, such that for every function $f(t)$ in the set, and every t,

$$|f(t+\tau_\varepsilon) - f(t)| \leq \varepsilon. \tag{14.03}$$

Hence any continuous function of a finite number of almost periodic functions yields an almost periodic function, as for example the sum or the product of a finite number of almost periodic functions. The limit of a uniformly convergent series or sequence of almost periodic functions is almost periodic. Every function that is periodic in the classical sense is almost periodic, and the same is true of

$$\Sigma A_n e^{i\Lambda_n t}, \tag{14.04}$$

in case $\Sigma |A_n|$ converges. Every almost periodic function is uniformly continuous. If $f(t)$ is almost periodic,

$$M\{f\} = \lim_{T\to\infty} \frac{1}{T} \int_{z}^{z+T} f(t)\,dt \tag{14.05}$$

exists as a uniform limit in z. If $f(t)$ is almost periodic, so is

$$\varphi(t) = M_x\{f(x+t)\bar{f}(x)\}. \tag{14.06}$$

(Here and later the symbol under the M indicates the variable on which the averaging is being done.) If $f(t)$ is a real non-negative almost periodic function, and $M\{f\}=0$, $f(t)$ is identically zero.

If $f(t)$ is almost periodic, then since $\varphi(t)$ is also almost periodic, it is continuous. Let us form

$$S(u)=\frac{1}{2\pi}\int_{-\infty}^{\infty}\varphi(x)\frac{e^{iux}-1}{ix}\,dx. \tag{14.07}$$

From theorems already established

$$\varphi(0)=S(\infty)-S(-\infty). \tag{14.08}$$

Let the discontinuities of $S(u)$ be at $u=\lambda_1, \lambda_2, \ldots$ These form a denumerable set, as $S(u)$ is of limited total variation, and indeed monotone. Let

$$a_n=S(\lambda_n+0)-S(\lambda_n-0). \tag{14.09}$$

All these coefficients a_n are positive, and

$$\sum_1^\infty a_n \leq S(\infty)-S(-\infty)=M\{|f|^2\}. \tag{14.10}$$

Let us form the function

$$\gamma(t)=\varphi(t)-\sum_0^\infty a_k e^{-i\lambda_k t}, \tag{14.11}$$

As a simple computation will show (cf. (4.05)),

$$\frac{1}{2\pi}\int_{-\infty}^{\infty}\gamma(x)\frac{e^{iux}-1}{ix}\,dx=S(u)-S_1(u), \tag{14.12}$$

where $S_1(u)$ consists of the sum of all the jumps of $S(u)$ with abscissae less than u, together with half the jump (if any) with abscissa u. Hence

$$S_2(u)=\frac{1}{2\pi}\int_{-\infty}^{\infty}\gamma(x)\frac{e^{iux}-1}{ix}\,dx \tag{14.13}$$

is a continuous function of limited total variation, say V. Let the total variation of $S_2(u)$ over the ranges $(-\infty, -B)$ and (B, ∞) be $V(B)$.

We have

$$\frac{1}{2\varepsilon}\int_{-\infty}^{\infty} |S_2(u+\varepsilon)-S_2(u-\varepsilon)|^2\,du$$

$$=\frac{1}{2\varepsilon}\left[\int_A^{\infty}+\int_{-\infty}^{-A}\right]|S_2(u+\varepsilon)-S_2(u-\varepsilon)|^2\,du+\frac{1}{2\varepsilon}\int_{-A}^{A}|S_2(u+\varepsilon)-S_2(u-\varepsilon)|^2\,du$$

$$\leq\frac{\max|S_2(v)|}{\varepsilon}\left[\int_A^{\infty}+\int_{-\infty}^{-A}\right]|S_2(u+\varepsilon)-S_2(u-\varepsilon)|\,du$$

$$+\max_{-A\leq v\leq A}|S_2(v+\varepsilon)-S_2(v-\varepsilon)|\frac{1}{2\varepsilon}\int_{-A}^{A}|S_2(u+\varepsilon)-S_2(u-\varepsilon)|\,du$$

$$\leq\frac{\max|S_2(v)|}{\varepsilon}\left[\int_{A-\varepsilon}^{A+\varepsilon}+\int_{A+\varepsilon}^{A+3\varepsilon}+\int_{A+3\varepsilon}^{A+5\varepsilon}+\cdots+\int_{-A-\varepsilon}^{-A+\varepsilon}+\int_{-A-3\varepsilon}^{-A-\varepsilon}\right.$$

$$\left.+\int_{-A-5\varepsilon}^{-A-3\varepsilon}+\cdots\right]|S_2(u+\varepsilon)-S_2(u-\varepsilon)|\,du+\max_{-A\leq v\leq A}|S_2(v+\varepsilon)-S_2(v-\varepsilon)|\frac{1}{2\varepsilon}$$

$$\left[\int_{-A-\varepsilon}^{-A+\varepsilon}+\int_{-A+\varepsilon}^{-A+3\varepsilon}+\cdots+\int_{-A+\left(\left[\frac{2A}{\varepsilon}\right]-2\right)\varepsilon}^{-A+\left[\frac{2A}{\varepsilon}\right]\varepsilon}\right]|S_2(u+\varepsilon)-S_2(u-\varepsilon)|\,du$$

$$=\frac{\max|S_2(v)|}{\varepsilon}\int_{-\varepsilon}^{\varepsilon}\{|S_2(u+A+\varepsilon)-S_2(u+A-\varepsilon)|+|S_2(u+A+3\varepsilon)$$

$$-S_2(u+A+\varepsilon)|+\cdots+|S_2(u-A+\varepsilon)-S_2(u-A+\varepsilon)|+|S_2(u-A-\varepsilon)$$

$$-S_2(u-A-3\varepsilon)|+\cdots\}\,du+\max_{-A\leq v\leq A}|S_2(v+\varepsilon)-S_2(v-\varepsilon)|$$

$$\cdot\frac{1}{2\varepsilon}\int_{-\varepsilon}^{\varepsilon}\Big\{|S_2(u-A+\varepsilon)-S_2(u-A-\varepsilon)|+|S_2(u-A+3\varepsilon)-S_2(u-A+\varepsilon)|$$

$$+\cdots+\left|S_2\left(u-A+\left[\frac{2A}{\varepsilon}\right]\varepsilon\right)-S_2\left(u-A+\left[\frac{2A}{\varepsilon}\right]\varepsilon-2\varepsilon\right)\right|\Big\}du$$

$$\leq 2\,V(A-2\,\varepsilon)\max|S_2(v)|+V\max_{-A\leq v\leq A}|S_2(v+\varepsilon)-S_2(v-\varepsilon)|. \tag{14. 14}$$

Since the function $S(u)$ is uniformly continuous over any finite range, this gives us

$$\lim_{\varepsilon\to 0}\frac{1}{2\,\varepsilon}\int_{-\infty}^{\infty}|S_2(u+\varepsilon)-S_2(u-\varepsilon)|^2\,du\leq 2\,V(A-\eta)\max|S_2(v)|. \tag{14. 15}$$

However, $V(A-\eta)$ tends to zero as A tends to infinity, and may be arbitrarily small. Hence

$$\lim_{\varepsilon\to 0}\frac{1}{2\,\varepsilon}\int_{-\infty}^{\infty}|S_2(u+\varepsilon)-S_2(u-\varepsilon)|^2\,du=0. \tag{14. 16}$$

Applying (5. 53), we get

$$\lim_{T\to\infty}\frac{1}{2\,T}\int_{-T}^{T}|\gamma(t)|^2\,dt=0. \tag{14. 17}$$

Since, however, $\gamma(t)$ and $|\gamma(t)|^2$ are almost periodic, we must have

$$\gamma(t)=0, \tag{14. 18}$$

which yields us

$$S(u)=S_1(u) \tag{14. 19}$$

and

$$\varphi(t)=\sum_{1}^{\infty}a_k e^{-i\lambda_k t}. \tag{14. 20}$$

Thus $S(u)$ is a step function, and the spectrum of an almost periodic function is a pure line spectrum.

15. The Parseval theorem for almost periodic functions.

A further result is

$$M\{\varphi(t)\,e^{i\lambda_k t}\}=a_k. \tag{15. 01}$$

If we remember the uniformity properties of the means of almost periodic functions, this yields

$$a_k = \lim_{T\to\infty} \frac{1}{2T}\int_{-T}^{T} e^{i\lambda_k t}\,dt \lim_{U\to\infty}\frac{1}{2U}\int_{-U}^{U} f(x+t)\bar{f}(x)\,dx$$

$$= \lim_{U\to\infty}\frac{1}{2U}\int_{-U}^{U}\bar{f}(x)e^{-i\lambda_k x}\,dx \lim_{T\to\infty}\frac{1}{2T}\int_{-T}^{T} f(x+t)e^{i\lambda_k(x+t)}\,dt$$

$$= \lim_{U\to\infty}\frac{1}{2U}\int_{-U}^{U}\bar{f}(x)e^{-i\lambda_k x}\,dx \lim_{T\to\infty}\frac{1}{2T}\int_{-T}^{T} f(y)e^{i\lambda_k y}\,dy$$

$$= |M\{f(x)e^{i\lambda_k x}\}|^2. \tag{15.02}$$

Hence

$$M\{|f|^2\} = \varphi(0) = S(\infty) - S(-\infty)$$

$$= S_1(\infty) - S_1(-\infty)$$

$$= \sum_0^\infty |M\{f(x)\,e^{i\lambda_k x}\}|^2. \tag{15.03}$$

This is a precise analogue to the Hurwitz-Parseval theorem for periodic functions, and is the well-known fundamental theorem of Bohr.

16. The Weierstrass theorem for almost periodic functions.

The present section 16 is devoted to the proof of the approximation theorem for almost periodic functions, which tells us that the necessary and sufficient condition for a function to be almost periodic is that it can be expressed as the uniform limit of a trigonometrical polynomial. The main idea of the present proof is due to Weyl, although the form of the argument is much changed from that on his paper. The essence of the proof is that harmonic analysis is not applied directly to the almost periodic function discussed, but to certain related functions derived from what Bochner calls the Verschiebungsfunktion of the given function. In the discussion of the many different extensions of almost periodic functions, there is a function in each case analogous to this Verschiebungsfunktion which is almost periodic in the strict Bohr sense. As we shall see in the next section, this enables us to carry over to these more general cases practically the entire Bohr approximation theorem redefined to suit each particular definition.

Let $f(t)$ be almost periodic. Consider

$$g(x) = \max_t |f(x+t) - f(t)|. \tag{16.01}$$

We have

$$\begin{aligned} |g(x+\tau) - g(x)| &\leq \max_t ||f(x+t+\tau) - f(t)| - |f(x+t) - f(t)|| \\ &\leq \max_t |f(x+t+\tau) - f(x+t)| \\ &= \max_t |f(t+\tau) - f(t)|. \end{aligned} \tag{16.02}$$

Hence any translation number for $f(t)$ pertaining to ε is a translation number for $g(t)$ pertaining to ε, and $g(t)$ is almost periodic. It is this function which Bochner calls the Verschiebungsfunktion of $f(t)$.

We have already indicated the fact that any continuous function of an almost periodic function is almost periodic. Let $H_\varepsilon(U)$ be befined as follows:

$$H_\varepsilon(U) = \begin{cases} 1; & [0 \leq U \leq \varepsilon/2] \\ 2 - \dfrac{2U}{\varepsilon}; & [\varepsilon/2 \leq U \leq \varepsilon] \\ 0. & [\varepsilon \leq U \quad] \end{cases} \tag{16.03}$$

Let

$$\psi_\varepsilon(x) = \frac{H_\varepsilon[g(x)]}{M_x H_\varepsilon[g(x)]}. \tag{16.04}$$

Since $H_\varepsilon[g(x)]$ is somewhere positive, and it is everywhere non negative and almost periodic, $M_x H_\varepsilon[g(x)]$ cannot vanish. Hence $\psi_\varepsilon(x)$ exists and is almost periodic.

Let

$$f_\varepsilon(x) = M_t\{f(t)\,\psi_\varepsilon(x-t)\}. \tag{16.05}$$

The existence and almost periodic character of $f_\varepsilon(x)$ are proved without difficulty. The definition of ψ_ε ensures that

$$|f(x) - f(t)| \leq \varepsilon \tag{16.06}$$

if $\psi_\varepsilon(x-t) \neq 0$. Hence, since $f_\varepsilon(x)$ is a mean of these values of $f(t)$,

$$\max_x |f(x) - f_\varepsilon(x)| \leq \varepsilon. \tag{16.07}$$

Similarly, if

$$f^{(\varepsilon)}(x) = M_t\{f_\varepsilon(t)\,\psi_\varepsilon(x-t)\}, \tag{16.08}$$

$f^{(\varepsilon)}(x)$ exists and is almost periodic, and

$$\max_x |f^{(\varepsilon)}(x) - f_\varepsilon(x)| \leq \varepsilon. \tag{16.09}$$

Hence, by (16.07) and (16.09),

$$\max_x |f^{(\varepsilon)}(x) - f(x)| \leq 2\,\varepsilon. \tag{16.10}$$

We have

$$f^{(\varepsilon)}(x) = M_t\{\psi_\varepsilon(x-t)\, M_\tau\{f(\tau)\,\psi_\varepsilon(t-\tau)\}\}. \tag{16.11}$$

Bearing in mind the uniformity properties of almost-periodic functions, we have

$$f^{(\varepsilon)}(x) = M_\tau\{f(\tau)\, M_t\{\psi_\varepsilon(x-t)\,\psi_\varepsilon(t-\tau)\}\}. \tag{16.12}$$

However, by (14.20),

$$M_t\{\psi_\varepsilon(x-t)\,\psi_\varepsilon(t-\tau)\} = \sum_1^\infty a_k e^{i\lambda_k(x-\tau)} \tag{16.13}$$

where all the coefficients a_k are positive, and $\sum_1^\infty a_k$ converges. Hence

$$f^{(\varepsilon)}(x) = M_\tau\left\{f(\tau)\sum_1^\infty a_k e^{i\lambda_k(x-\tau)}\right\}. \tag{16.14}$$

Since

$$\left|\frac{1}{2T}\int_{-T}^{T} f(\tau)\, e^{i\lambda_k(x-\tau)}\, d\tau\right| \leq \max|f(\tau)|, \tag{16.15}$$

it follows that we can invert the order of M and Σ, and that

$$f^{(\varepsilon)}(x) = \sum_1^\infty a_k e^{i\lambda_k x}\, M\{f(\tau)\, e^{-i\lambda_k\tau}\}. \tag{16.16}$$

Inasmuch as

$$|M\{f(\tau)\, e^{-i\lambda_k\tau}\}| \leq \max|f(\tau)|, \tag{16.17}$$

$f^{(\varepsilon)}(x)$ is the sum of a uniformly convergent series of trigonometric terms. That is to say, we can choose N so large that

31—29764. *Acta mathematica.* 55. Imprimé le 29 avril 1930.

$$\max_x \left| f^{(\varepsilon)}(x) - \sum_1^N a_k e^{i\lambda_k x} M\{f(\tau)e^{-i\lambda_k \tau}\} \right| \leq \varepsilon, \tag{16.18}$$

and hence that

$$\max_x \left| f(x) - \sum_1^N a_k e^{i\lambda_k x} M\{f(\tau)e^{-i\lambda_k \tau}\} \right| \leq 3\,\varepsilon. \tag{16.19}$$

In other words, we have proved Bohr's approximation theorem, to the effect that it is possible to approximate uniformly to any desired degree of accuracy to an almost periodic function by means of trigonometrical polynomials.

17. Certain generalizations of almost periodic functions.

It will be noticed that in the proof of the Weierstrass theorem for almost periodic functions, the spectrum of the function to be analyzed was not directly introduced, but rather that of the auxiliary function $\psi_\varepsilon(t)$. In many cases, when the function $f(t)$ is not almost periodic in the classical sense, an auxiliary function $\psi_\varepsilon(t)$ may be defined, which will be almost periodic in the classical sense, and which may be employed to establish the approximation theorem for $f(t)$, in whatever sense this theorem may hold. It would be possible in this manner to establish the approximation theorems for the almost periodic functions of the generalized types of Weyl, Besicovitch, Stepanoff, and others, but one example will suffice to show the power of the method, and to this we shall confine ourselves. This example, which is due to Mr. C. F. Muckenhoupt, is that of functions almost periodic in the mean.

We shall confine our attention to functions $f(x, t)$ defined over the range $(-\infty < t < \infty,\ x_0 \leq x \leq x_1)$, quadratically summable in x, and continuous in the mean in t in the sense that

$$\lim_{\tau \to 0} \int_{x_0}^{x_1} |f(x, t) - f(x, t+\tau)|^2 dx = 0. \tag{17.01}$$

We shall say that τ_ε is a *translation number* of $f(x, t)$ pertaining to ε in case for all t,

$$\int_{x_0}^{x_1} |f(x, t+\tau_\varepsilon) - f(x, t)|^2 dx < \varepsilon^2. \tag{17.02}$$

In case, given ε, we can always assign a finite quantity L_ε, such that each interval $(A, A+L_\varepsilon)$ contains at least one translation number τ_ε pertaining to ε, then $f(x, t)$ is said to be *almost periodic in the mean.* I case $f(x, t)$ is almost periodic in the mean, Mr. Muckenhoupt's theorem is:

Given any positive quantity ε there can be assigned a trigonometrical polynomial

$$P_\varepsilon(x, t) = \sum_1^N A_n(x) e^{i\Lambda_n t}, \tag{17.03}$$

such that

$$A_1(x),\ A_2(x),\ \ldots,\ A_n(x) \tag{17.04}$$

are all quadratically summable, and for all t,

$$\int_{x_0}^{x_1} |f(x, t) - P_\varepsilon(x, t)|^2 \, dx < \varepsilon^2. \tag{17.05}$$

There are a number of elementary theorems which Mr. Muckenhoupt proves along lines not differing in any essential way from those followed by Bohr in the proof of the corresponding theorems for functions almost periodic in the original sense. Thus every function almost periodic in the mean is bounded in the mean, in the sense that

$$\int_{x_0}^{x_1} |f(x, t)|^2 \, dx \tag{17.06}$$

is bounded; and is uniformly continuous in the mean, in the sense that

$$\lim_{\varepsilon \to 0} \max_{\tau < \varepsilon} \int_{x_0}^{x_1} |f(x, t) - f(x, t+\tau)|^2 \, dx = 0. \tag{17.07}$$

In this and subsequent formulas, the maximum value indicated by »max» need not be actually attained. Any finite set of functions almost periodic in the mean are *simultaneously* almost periodic in the mean, in the sense that, given ε, an L_ε may be assigned in such a manner that every interval $(A, A+L_\varepsilon)$ contains at least one τ_ε which is a translation number pertaining to ε of all the functions of the set. Hence the sum of two or more functions almost periodic in the mean is almost periodic in the mean. Similarly, the product in the ordinary sense is almost periodic in the mean. The uniform limit in the mean of a set of functions almost periodic in the mean is itself almost periodic in the mean. If $f(x, t)$ is almost periodic in the mean,

$$\underset{T\to\infty}{\text{l.i.m.}}\ \frac{1}{T}\int_{a}^{a+T} f(x, t)\, dt \tag{17.08}$$

exists as a uniform limit in the mean in a, and is independent of a. We shall represent it by the symbol

$$M\{f(x, t)\}. \tag{17.09}$$

Mr. Muckenhoupt now puts

$$g(t) = \max_{z} \int_{x_0}^{x_1} |f(x, t+z) - f(x, z)|^2\, dx. \tag{17.10}$$

Clearly

$$\begin{aligned} |g(t+\tau) - g(t)| &\leq \max_{z} \int_{x_0}^{x_1} \big||f(x, t+\tau+z) - f(x, z)|^2 - |f(x, t+z) - f(x, z)|^2\big|\, dx \\ &\leq \max_{z} \left\{ \int_{x_0}^{x_1} \big||f(x, t+\tau+z) - f(x, z)| + |f(x, t+z) - f(x, z)|\big|^2\, dx \right\}^{1/2} \\ &\quad \cdot \max_{z} \left\{ \int_{x_0}^{x_1} \big||f(x, t+\tau+z) - f(x, z)| - |f(x, t+z) - f(x, z)|\big|^2\, dx \right\}^{1/2}. \end{aligned} \tag{17.11}$$

To evaluate this, let us consider the maximum of each of the integrals under the radical sign separately. The first does not exceed

$$16 \max_{z} \int_{x_0}^{x_1} |f(x, z)|^2\, dx; \tag{17.12}$$

the second does not exceed

$$\max_{z} \int_{x_0}^{x_1} |f(x, \tau+z) - f(x, z)|^2\, dx. \tag{17.13}$$

Hence we may write

$$|g(t+\tau)-g(t)| \leq 16\left[\max_z \int_{x_0}^{x_1} |f(x, z+\tau)-f(x, z)|^2\, dx\right]^{1/2}$$

$$\cdot\left[\max_z \int_{x_0}^{x_1} |f(x, z)|^2 dx\right]^{1/2}. \qquad (17.\ 14)$$

If τ_ε is a translation number of $f(x, t)$ pertaining to ε, we have

$$|g(t+\tau_\varepsilon)-g(t)| \leq 16\,\varepsilon\left[\max_z \int_{x_0}^{x_1} |f(x, z)|^2 dx\right]^{1/2}, \qquad (17.\ 15)$$

so that any translation number of $f(x, t)$ pertaining to ε is a translation number of $g(t)$ pertaining to

$$16\,\varepsilon\left[\max_z \int_{x_0}^{x_1} |f(x, z)|^2 dx\right]^{1/2}. \qquad (17.\ 16)$$

Thus $g(t)$ is almost periodic in the classical sense, and is distinct from 0 unless $f(x, t)$ is independent of t, in the sense that $f(x, t_1)=f(x, t_2)$ almost everywhere.

As in the last section, let

$$H_\varepsilon(U) = \begin{cases} 1 \quad ; & [0 \leq U \leq \varepsilon/2] \\ 2-\dfrac{2U}{\varepsilon}; & [\varepsilon/2 \leq U \leq \varepsilon] \\ 0 \quad ; & [\varepsilon \leq U \quad] \end{cases} \qquad (17.\ 17)$$

and let

$$\psi_\varepsilon(t) = \frac{H_\varepsilon[g(t)]}{M H_\varepsilon[g(t)]}. \qquad (17.\ 18)$$

As before, $H_\varepsilon[g(t)]$ is dictinct from 0, and $\psi_\varepsilon(x)$ exists, and is almost periodic. As before, we put

$$f_\varepsilon(x, t) = M_\tau[f(x, \tau)\,\psi_\varepsilon(t-\tau)], \qquad (17.\ 19)$$

and

$$f^{(\varepsilon)}(x, t) = M_\tau[f(x, \tau)\,\psi_\varepsilon(t-\tau)]. \qquad (17.\ 20)$$

A proof precisely parallel to that of (16. 10) and (16. 12) shows that

$$\max_t \int_{x_0}^{x_1} |f(x, t) - f^{(\varepsilon)}(x, t)|^2 \, dx \leq 4\,\varepsilon. \tag{17.21}$$

and that

$$f^{(\varepsilon)}(x, t) = M_\sigma[f(x, \sigma)\, M_\tau[\psi_\varepsilon(t-\tau)\, \psi_\varepsilon(\tau-\sigma)]]. \tag{17.22}$$

As before,

$$M_\tau[\psi_\varepsilon(t-\tau)\, \psi_\varepsilon(\tau-\sigma)] = \sum_0^\infty a_k e^{i\lambda_k(t-\sigma)} \tag{17.23}$$

where all the a_k's are positive, and $\sum_0^\infty a_k$ converges. Hence

$$f^{(\varepsilon)}(x, t) = M_\sigma\left[f(x, \sigma) \sum_0^\infty a_k e^{i\lambda_k(t-\sigma)}\right]. \tag{17.24}$$

We have

$$\lim_{N\to\infty} \int_{x_0}^{x_1} \left| M_\tau \left[f(x, \sigma) \sum_N^\infty a_k e^{i\lambda_k(t-\sigma)}\right]\right|^2 dx$$

$$\leq \lim_{N\to\infty} \left[\sum_N^\infty a_k\right]^2 \max_\sigma \int_{x_0}^{x_1} |f(x, \sigma)|^2 \, dx = 0. \tag{17.25}$$

Hence since

$$\sum_0^\infty a_k$$

converges, we can invert the order of M and Σ in (17.25), and get

$$f^{(\varepsilon)}(x, t) = \underset{N\to\infty}{\mathrm{l.i.m.}} \sum_0^n a_k e^{i\lambda_k t} \underset{\sigma}{M}[f(x, \sigma)\, e^{-i\lambda_k \sigma}]. \tag{17.26}$$

This convergence in the mean is uniform with respect to t. Combining (17.25) and (17.26), our theorem is proved.

This theorem has an interesting dynamical application. Really significant dynamical applications of almost periodic functions have been rather scarce, as no one has yet produced an example of an almost periodic function entering into a dynamical system with a finite number of degrees of freedom in which the frequencies or exponents are not linearly dependent (with rational coefficients) on a finite set of quantities. However, dynamical systems with an infinite number of degress of freedom are familiar enough in connection with boundary value problems, and in these, it is well known that the solution may involve an infinite linearly independent set of time frequencies. Mr. Muckenhoupt has succeeded in showing, under certain very general conditions, that the solution of such a problem is almost periodic in the mean with respect to the time, the space variables playing the rôle above assigned to x. In this proof, the existence of an integral invariant such as the energy is of the utmost importance, as is also the condition that when all the coordinates and velocities of the system are less in value than some given constant, the energy is also necessarily less than some constant.

Let us consider as an example a vibrating string, whose density and tensions are functions of position, but not of time. Let the mass density be $\mu(x)$ and the tension $T(x)$. The equation of motion is then

$$\frac{\partial}{\partial x}\left[T(x)\frac{\partial y}{\partial x}\right]=\mu(x)\frac{\partial^2 y}{\partial t^2}. \tag{17.27}$$

We consider the ends to be fixed, giving us

$$y(x_0)=y(x_1)=0, \tag{17.28}$$

and we take T and μ, as is always physically the case, finite and positive. We shall also suppose them to have bounded derivatives of all orders.

Thus the total energy of the system is

$$E_0=\frac{1}{2}\int_{x_0}^{x_1}\left[\mu(x)\left(\frac{\partial y}{\partial t}\right)^2+T(x)\left(\frac{\partial y}{\partial x}\right)^2\right]dx. \tag{17.29}$$

If we assume density and tension independent of the time, we have

$$\frac{\partial E_0}{\partial t}=\int_{x_0}^{x_1}\left[\mu(x)\frac{\partial y}{\partial t}\frac{\partial^2 y}{\partial t^2}+T(x)\frac{\partial y}{\partial x}\frac{\partial^2 y}{\partial x\,\partial t}\right]dx, \tag{17.30}$$

or by (17.27),

$$\frac{\partial E_0}{\partial t}=\int_{x_0}^{x_1}\left\{\frac{\partial y}{\partial t}\frac{\partial}{\partial x}\left[T(x)\frac{\partial y}{\partial x}\right]+T(x)\frac{\partial y}{\partial x}\frac{\partial}{\partial x}\left(\frac{\partial y}{\partial t}\right)\right\}dx$$

$$=\left[T(x)\frac{\partial y}{\partial x}\frac{\partial y}{\partial t}\right]_{x_0}^{x_1}=0. \tag{17.31}$$

Thus the total energy is invariant, as was to be expected from physical considerations.

Inasmuch as $\partial^n y/\partial t^n$ also satisfies equation (17. 27) and boundary conditions (17. 28), we see that all the expressions

$$E_n = \frac{1}{2}\int_{x_0}^{x_1}\left[\mu(x)\left(\frac{\partial^{n+1}y}{\partial t^{n+1}}\right)^2 + T(x)\left(\frac{\partial^{n+1}y}{\partial t\,\partial x^n}\right)^2\right]dx \tag{17. 32}$$

are invariants, at least if $y(x, t)$ is sufficiently often differentiable. We shall term E_n the $(n+1)$st energy of the system.

Let us now take $\partial y/\partial t$ and $\partial y/\partial x$ to be continuous, and let

$$E_0 \leq E; \quad \mu(x) \geq M; \quad T(x) \geq T.$$

Clearly $E_0 \geq 0$, and

$$\int_{x_0}^{x_1}\left(\frac{\partial y}{\partial t}\right)^2 dx \leq \frac{2\,E_0}{M} \leq \frac{2\,E}{M}. \tag{17. 33}$$

Similarly, $\int_{x_0}^{x_1}\left(\frac{\partial y}{\partial x}\right)^2 dx$ is bounded, provided only the first energy E_0 is finite. Furthermore, since

$$y^2 = \left[\int_{x_0}^{x}\frac{\partial y}{\partial x}\,dx\right]^2 \leq \int_{x_0}^{x}dx\int_{x_0}^{x}\left(\frac{\partial y}{\partial x}\right)^2 dx, \tag{17. 34}$$

by the Schwarz inequality, y is bounded. Similarly, if E_1 is also bounded,

$$\int_{x_0}^{x_1}\left(\frac{\partial^2 y}{\partial t^2}\right)^2 dx, \quad \int_{x_0}^{x_1}\left(\frac{\partial^2 y}{\partial x\,\partial t}\right)^2 dx, \quad \text{and} \quad \int_{x_0}^{x_1}\left(\frac{\partial^2 y}{\partial x^2}\right)^2 dx \tag{17. 35}$$

will be likewise; in the last case, as a result of (17. 27); if E_2 is also bounded,

$$\int_{x_0}^{x_1}\left(\frac{\partial^3 y}{\partial t^3}\right)^2 dx, \quad \int_{x_0}^{x_1}\left(\frac{\partial^3 y}{\partial t^2\,\partial x}\right)^2 dx, \quad \int_{x_0}^{x_1}\left(\frac{\partial^3 y}{\partial t\,\partial x^2}\right)^2 dx, \quad \text{and} \quad \int_{x_0}^{x_1}\left(\frac{\partial^3 y}{\partial x^3}\right)^2 dx \tag{17. 36}$$

will be, and so on indefinitely.

Let us now introduce

$$(y, y_1) = \left\{ \int_{x_0}^{x_1} \left[|y - y_1|^2 + \left| \frac{\partial y}{\partial x} - \frac{\partial y_1}{\partial x} \right|^2 + \left| \frac{\partial y}{\partial t} - \frac{\partial y_1}{\partial t} \right|^2 \right] dx \right\}^{1/2} \tag{17.37}$$

as the distance between two functions, $y(x, t)$ and $y_1(x, t)$. If we write

$$y \sim \sum_{-\infty}^{\infty} A_n(t) e^{\frac{2\pi i n x}{x_1 - x_0}}; \quad y_1 \sim \sum_{-\infty}^{\infty} B_n(t) e^{\frac{2\pi i n x}{x_1 - x_0}};$$

we may approximate *uniformly* to (y, y_1) by

$$\int_{x_0}^{x_1} \left[\left| \sum_{-N}^{N} (A_n - B_n) e^{\frac{2\pi i n x}{x_1 - x_0}} \right|^2 \left(1 + \frac{4\pi n^2}{(x_1 - x_0)^2} \right) + \left| \sum_{-N}^{N} (A'_n - B'_n) e^{\frac{2\pi i n x}{x_1 - x_0}} \right|^2 dx \right] \tag{17.38}$$

for all functions y and y_1 for which E_0 and E_1 are finite, since then

$$\left. \begin{aligned} \sum_{-\infty}^{\infty} n^4 |A_n(t) - B_n(t)|^2 &= \frac{x_1 - x_0}{(2\pi)^4} \int_{x_0}^{x_1} \left(\frac{\partial^2 y}{\partial t^2} \right)^2 dx; \\ \sum_{-\infty}^{\infty} n^2 |A'_n(t) - B'_n(t)|^2 &= \frac{x_1 - x_0}{(2\pi)^2} \int_{x_0}^{x_1} \left(\frac{\partial^2 y}{\partial x \partial t} \right)^2 dx \end{aligned} \right\} \tag{17.39}$$

are uniformly bounded. Now, a bounded region in space of m dimensions may be divided into a finite number of compartments such that the distance between two points in the same compartment does not exceed ε. Hence we can divide the entire class of functions $y(x, t)$ for which E_0 and E_1 are finite into a finite number of classes such that the distance between two functions in the same class does not exceed ε.

Let us do this, and let us discard every class which is not actually represented by $y(x, t)$ for some value of t. Then we may assign a time-interval L'_ε within which $y(x, t)$ enters every class that it ever enters. Then, whatever τ may be, we may determine τ_1 between 0 and L'_ε such that

$$(y(x, \tau), \quad y(x, \tau_1))^2 < \varepsilon. \tag{17.40}$$

Since

$$y(x, t) - y(x, t + \tau - \tau_1)$$

satisfies the differential equation (17. 27),

$$\mathfrak{E} = \frac{1}{2}\int_{x_0}^{x_1}\left[\mu(x)\left(\frac{\partial y(x,t)}{\partial t} - \frac{\partial y(x,t+\tau-\tau_1)}{\partial t}\right)^2 + T(x)\left(\frac{\partial y(x,t)}{\partial x} - \frac{\partial y(x,t+\tau-\tau_1)}{\partial x}\right)^2\right] dx \tag{17. 41}$$

is invariant, and since for $t = \tau_1$,

$$\begin{aligned}\mathfrak{E} &\leq (\max \mu + \max T)\,(y(x,\tau_1),\ y(x,\tau))^2 \\ &< (\max \mu + \max T)\,\varepsilon,\end{aligned} \tag{17. 42}$$

it follows that for all t,

$$\mathfrak{E} < (\max \mu + \max T)\varepsilon, \tag{17. 43}$$

and hence by (17. 34) and (17. 37)

$$|y(x,t) - y(x,t+\tau-\tau_1)| < \sqrt{2\,\frac{x_1 - x_0}{M}(\max \mu + \max T)\varepsilon}\,. \tag{17. 44}$$

Since for every τ, there is a value of τ_1 between 0 and L'_ε, there is a value of $\tau - \tau_1$ over every interval of length L'_ε. Thus $y(x,t)$ is an a almost periodic function taken with respect to the time, uniformly in x, and is *a fortiori* almost periodic in the mean, in case E_0 and E_1 are finite. It follows that we may so determine $A_1(x), \ldots, A_n(x)$; $\Lambda_1, \ldots, \Lambda_n$ that for all t,

$$\int_{x_0}^{x_1}\left|y(x,t) - \sum_1^n A_k(x)\,e^{i\Lambda_k(t)}\right|^2 dx < \varepsilon. \tag{17. 45}$$

It is possible to go further than this, as Mr. Muckenhoupt has done, and to show that the method we have given for obtaining $f_\varepsilon(x,t)$, $f^{(\varepsilon)}(x,t)$, and $A_k(x)\,e^{i\Lambda_k(t)}$ assures us that all the functions

$$A_k(x)\,e^{i\Lambda_k t}$$

are solutions of the original differential equation, or that the functions

$$A_k(x)$$

are all solutions of the ordinary differential equation

$$\frac{d}{dx}(T(x)\,A'_k(x)) + \Lambda_k^2\,\mu(x)\,A_k(x) = 0; \tag{17. 46}$$

— that is, are what is known as Eigenfunktionen of the dynamical problem. This proof rests on the fact that each one of these functions may be obtained from its predecessor, and ultimately from $f(x, t)$, by a process of weighted averaging in the variable t which transforms every solution of a linear differential equation with coefficients constant with respect to the time into another solution of the same equation, or at least of the corresponding integral equation. Hence, if $y(x, 0) = F(x); \frac{\partial y(x, 0)}{\partial t} = 0$ is a possible set of initial conditions for the motion of the vibrating string. We may write

$$F(x) = \underset{n \to \infty}{\text{l.i.m.}} \sum_{1}^{n} A_k(x), \tag{17.47}$$

where the $A_k(x)$ are in general Eigenfunktionen of the problem that depend on n. Thus if the set of possible initial conditions of the string is closed, as we may show to be the case by direct methods, every quadratically summable function may be expanded in terms of a denumerable set of Eigenfunktionen, and the Eigenfunktionen may be shown to be a denumerable closed set.

The methods of Mr. Muckenhoupt are susceptible of extension to the treatment of a much wider class of Eigenfunktion problems, in any finite number of dimensions. The detail of this extension awaits further investigation.

Bibliography.

The works and papers covering the various theories belonging to general harmonic analysis fall into several imperfectly related categories. Among these are:

(1) The various papers written from the physical standpoint, with the explicit purpose of clearing up obscure points in the theories of interference, of coherency, and of polarization.

(2) Directly related to these, the various memoirs connecting with the Schuster theory of the periodogram.

(3) A group of memoirs preceding the Bohr theory of almost periodic functions, applying various extensions of the notion of periodicity in celestial mechanics and other similar fields.

(4) Papers written from the point of view of the mathematician, and dealing with trigonometric series not proceeding according to integral multiples of the argument.

(5) The Bohr theory of almost periodic functions, and papers directly inspired by it.

(6) Papers dealing with haphazard motion, and using ideas directly pertinent to generalized harmonic analysis.

(7) The Hahn direction of work, treating generalized harmonic analysis from the standpoint of ordinary convergence, rather than from that of convergence in the mean.

(8) The papers assuming essentially the standpoint of the present author, in whose work the generalizations of the Parseval theorem play the central rôle.

(9) Papers dealing rather with the rigorous theory of the Fourier integral itself than with its generalizations.

(10) Papers not dealing directly with generalized harmonic analysis, which it is desirable to cite for one reason or another.

In citing any paper, it will be indicated to which of these categories it belongs. Each paper will furthermore be quoted in the footnotes by the name of its author, together with an index number given in the bibliography.

A. C. BERRY 1. Doctoral dissertation, Harvard, 1929. Unpublished. (8)

—— 2. The Fourier transform theorem. Jour. Math. and Phys. Mass. Inst. Technology 8, 106—118. (1929). (8)

A. BESICOVITCH 1. Sur quelques points de la théorie des fonctions presque périodiques. C.R. 180, 394—397. (1925). (5)

—— 2. On Parseval's theorem for Dirichlet series. Proc. Lond. Math. Soc. 25, 25—34. (1926). (5)

—— 3. On generalized almost periodic functions. Proc. Lond. Math. Soc. 24, 495—512. (1926). (5)

A. BESICOVITCH and H. BOHR 1. Some remarks on generalizations of almost periodic functions. Danske Vidensk. Selskab. 8, No. 5. (1927). (5)

—— 2. On generalized almost periodic functions. Journ. Lond. Math. Soc. 3, 172—176. (1928). (5)

G. D. BIRKHOFF. Dynamical systems. Am. Math. Society, New York, 1927. Pp. 218—220. (5)

S. BOCHNER 1. Properties of Fourier series of almost periodic functions. Proc. Lond. Math. Soc. 26, 433. (1925). (5)

—— 2. Sur les fonctions presque-périodiques de Bohr. C.R. 180, 1156. (1925). (5)

—— 3. Beiträge zur Theorie der fastperiodischen Funktionen. Math. Ann. 96, 119—147. (1926). (5)

—— 4. Beiträge zur Theorie der fastperiodischen Funktionen. Math. Ann. 96, 383—409. (1926). (5)

—— 5. Über Fourierreihen fastperiodischer Funktionen. Berliner Sitzungsberichte, 26. (1926). (5)

—— 6. Konvergenzsätze für Fourierreihen grenzperiodischer Funktionen. Math. Zeitschr. 27, 187—211. (1927). (5)

—— 7. Darstellung reellvariabler und analytischer Funktionen durch verallgemeinerte Fourier- und Laplace-Integrale. Math. Ann. 97, 632—674. (1926—7). (8)

S. BOCHNER 8. Über gewisse Differential- und allgemeinere Gleichungen, deren Lösungen fastperiodisch sind. I. Teil. Der Existenzsatz. Math. Ann. 102, 489—504. (1929). (5)

S. BOCHNER and G. H. HARDY. Note on two theorems of Norbert Wiener. Jour. Lond. Math. Soc. 1, 240—242. (1926). (10)

P. BOHL. Über die Darstellung von Funktionen einer Variabeln durch trigonometrische Reihen mit mehreren einer Variabeln proportionalen Argumenten. (Dorpat, 1893). (3)

H. BOHR 1. Sur les fonctions presque périodiques. C.R. Oct. 22, 1923. (5)

—— 2. Sur l'approximation des fonctions presque périodiques par des sommes trigonométriques. C.R. Nov. 26, 1923. (5)

—— 3. Über eine quasi-periodische Eigenschaft Dirichletscher Reihen mit Anwendung auf die Dirichletschen L Funktionen. Math. Ann. 85, 115—122. (4)

—— 4. Zur Theorie der fastperiodischen Funktionen. Acta Math. 45, 29—127. (124). (5)

—— 5. Zur Theorie der fastperiodischen Funktionen. II. Acta Math. 46, 101—214. (1925). (5)

—— 6. Zur Theorie der fastperiodischen Funktionen. III. Acta Math. 47, 237—281. (1926). (5)

—— 7. Einige Sätze über Fourierreihen fastperiodischer Funktionen. Math. Ztschr. 23, 38—44. (1925). (5)

—— 8. Sur une classe de transcendantes entières. C.R. 181, 766. (1925). (5)

—— 9. Sur le théorème d'unicité dans la théorie des fonctions presque-périodiques. Bull. Sci. Math. 50, 1—7. (1926). (5)

—— 10. On the explicit determination of the upper limit of an almost periodic function. Jour. Lond. Math. Soc. 1. (1926). (5)

—— 11. Ein Satz über analytische Fortsetzung fastperiodischer Funktionen. Crelle, 157, 61—65. (5)

—— 12. En Klasse hele transcendente Funktioner. Mat. Tidsskrift, B. Aarg. 1926, 41—45. (5)

—— 13. Allgemeine Fourier- und Dirichlet-Entwicklungen. Abh. aus dem math. Sem. d. Hamburgischen Univ. 4, 366—374. (1926). (5)

—— 14. Fastperiodische Funktionen. Jahresb. d. D.M.V. 33, 25—41. (1925). (5)

—— 15. En Sætning om Fourierrækker for næstenperiodiske Funktioner. Mat. Tid. B. Aarg. 1924, 31—37. (5)

—— 16. Über die Verallgemeinerungen fastperiodischer Funktionen. Math. Ann. 99, 357—366. (1928). (5)

H. BOHR and O. NEUGEBAUER. Über lineare Differentialgleichungen mit konstanten Koeffizienten und fastperiodischer rechter Seite. Gött. Sitzb. 1926, 1—13. (5)

E. BOREL. Les probabilités dénombrables et leurs applications arithmétiques. Rend. di Palermo 27, 247—271. (1909). (10)

J. C. BURKILL 1. The expression in Stieltjes integrals of the inversion formulae of Fourier and Hankel. Proc. Lond. Math. Soc. 25, 513—524. (1926). (7)

J. C. BURKILL 2. On Mellin's inversion formula. Proc. Camb. Phil. Soc. 23, 356—360. (1926). (7)

V. BUSH, F. D. GAGE, and H. R. STEWART. A continuous integraph. Jour. Franklin Institute. 63—84. (1927). (10)

V. BUSH and H. L. HAZEN. Integraph solution of differential equations. 575—615. (1927). (10)

CARSE and SHEARER. A course in Fourier's analysis and periodogram analysis for the mathematical laboratory. London, 1915. (2)

P. J. DANIELL 1. A general form of integral. Annals of Math. 19, 279—294. (1918). (10)

—— 2. Integrals in an infinite number of dimensions. Ann. of Math. 20, 281—288. (1919). (10)

—— 3. Further properties of the general integral. Ann. of Math. 21, 203—220. (1920). (10)

W. DORN. Fouriersche Integrale als Grenzwerte Fourierscher Reihen. Wiener Sitzungsber., 1926, 127—147. (7)

A. EINSTEIN. Zur Theorie der Brownschen Bewegung. Ann. der Phys. (4) 19, 372—381. (1906). (10)

E. ESCLANGON 1. Sur une extension de la notion de périodicité. C.R. 135, 891—894. (1902). (3)

—— 2. Sur les fonctions quasi périodiques moyennes, déduites d'une fonction quasi périodique. C.R. 157, 1389—1392. (1913). (3)

—— 3. Sur les intégrales quasi périodiques d'une équation différentielle linéaire. C.R. 160, 652—653. (1915). (3)

—— 4. Sur les intégrales quasi périodiques d'une équation différentielle linéaire. C.R. 161, 488—489. (1915). (3)

J. FAVARD 1. Sur les fonctions harmoniques presque-périodiques. C.R. 182, 757. (1926). (5)

—— 2. Sur les equations différentielles à coefficients presque-periodiques. Acta Math. 51, 31—81. (1927). (5)

P. FRANKLIN 1. Almost periodic recurrent motions. Math. Ztschr. 30, 325—331. (1929). (5)

—— 2. The elementary theory of almost periodic functions of two variables. Jour. Math. Phys. M.I. T. 5, 40—55. (1925). (5)

—— 3. The fundamental theorem of almost periodic functions of two variables. Jur. Math. Phys. M.I. T. 5, 201—237. (1926). (5)

—— 4. Classes of functions orthogonal on an infinite interval, having the power of the continuum. Jour. Math. Phys. M.I. T. 8, 74—79. (1929). (5)

—— 5. Approximation theorems for generalized almost periodic functions. Math. Ztschr. 29, 70—87. (1928). (5)

G. L. GOUY. Sur le mouvement lumineux. Journal de physique, 5, 354—362. (1886). (1)

H. HAHN 1. Über die Verallgemeinerung der Fourierschen Integralformel. Acta Math. 49, 301—353. (1926). (7)

H. HAHN 2. Über die Methode der arithmetischen Mittel in der Theorie der verallgemeinerten Fourierintegrale. Wiener Sitzungsber., 1925, 449—470. (7)

G. H. HARDY, A. E. INGHAM, and G. PÓLYA. Notes on moduli and mean values. Proc. Lond. Math. Soc. 27, 401—409. (1928). (10)

E. W. HOBSON. The theory of functions of a real variable and the theory of Fourier series. Vol. 2, second edition. Cambridge, 1927. (9)

S. IZUMI 1. Über die Summierbarkeit der Fourierschen Integralformel. Tôhoku Journal 30, 96—110. (1929). (7)

—— 2. On the Cahen-Mellin's Inversion Formula. Tôhoku Journal 30, 111—114. (1929). (7)

M. JACOB 1. Über ein Theorem von Bochner-Hardy-Wiener. Jour. Lond. Math. Soc. 3, 182—187. (1928). (10)

—— 2. Über den Eindeutigkeitssatz in der Theorie der verallgemeinerten trigonometrischen Integrale. Math. Ann. 100, 279—294. (1928). (7)

—— 3. Über den Eindeutigkeitssatz in der Theorie der trigonometrischen Integrale. Math. Ann. 97, 663—674. (1927). (9)

G. W. KENRICK 1. Doctoral dissertation, Mass. Inst. Technology, 1927. (8)

—— 2. The analysis of irregular motions with applications to the energy-frequency spectrum of static and of telegraph signals. Phil. Mag. (7) 7, 176—196. (1929). (8)

E. H. LINFOOT. Generalization of two theorems of H. Bohr. Jour. Lond. Math. Soc. 3, 177—182. (1928). (5)

S. B. LITTAUER. On a theorem of Jacob. Jour. Lond. Math. Soc. 4, 226—231. (1929). (10)

K. MAHLER. On the translation properties of a simple class of arithmetical functions. J. Math. Phys. Mass. Inst. Technology 6, 158—164. (1927). (8)

C. F. MUCKENHOUPT. Almost periodic functions and vibrating systems. Doctoral dissertation, Mass. Inst. Technology, 1929. J. Math. Phys. Mass. Inst. Technology 8, 163—200. (1929). (5)

M. PLANCHEREL 1. Contribution à l'étude de la répresentation d'une fonction arbitraire par des intégrales définies. Rendiconti di Palermo, 30, 289—335. (1910). (9)

—— 2. Sur la représentation d'une fonction arbitraire par une intégrale définie. C.R. 150, 318—321. (1910). (9)

—— 3. Sur la convergence et sur la sommation par les moyennes de Cesàro de $\lim_{z=\infty} \int_a^z f(x) \cos xy \, dx$. Math. Ann. 76, 315—326. (1915). (9)

H. POINÇARÉ. Leçons sur la théorie mathématique de la lumière. Paris, 1889. (1)

S. POLLARD 1. The summation of a Fourier integral. Proc. Camb. Phil. Soc. 23, 373—382. (1926). (9)

—— 2. On Fourier's integral. Proc. Lond. Math. Soc. 26, 12—24. (1927). (9)

S. POLLARD 3. Identification of the coefficients in a trigonometrical integral. Proc. Lond. Math. Soc. 25, 451—468. (1926). (9)

A. PRINGSHEIM. Über neue Gültigkeitsbedingungen der Fourierschen Integralformel. Math. Ann. 68, 367—408. (1910). (9)

J. RADON, Theorie und Anwendungen der absolutadditiven Mengenfunktionen. Wien. Ber. 122, 1295—1438. (1913). (10)

Lord RAYLEIGH 1. On the resultant of a large number of vibrations of the same pitch and of arbitrary phase. Phil. Mag. 10, 73—78. (1880). (6)

—— 2. Wave theory of light. Ency. Britt., 1888. Cf. especially § 4. (1)

—— 3. On the character of the complete radiation at a given temperature. Phil. Mag. 27, 460. (1889). (1)

—— 4. Röntgen rays and ordinary light. Nature, 57, 607. (1898). (1)

—— 5. On the spectrum of an irregular disturbance. Phil. Mag. 5, 238—243. (1903). (6)

—— 6. Remarks concerning Fourier's theorem as applied to physical problems. Phil. Mag. 24, 864—869. (1912). (1)

—— 7. On the problem of random vibrations, and of random flights in one, two, or three dimensions. Phil. Mag. 37, 321—347. (1919). (6)

—— 8. On the resultant of a number of unit vibrations, whose phases are at random over a range not limited to an integral number of periods. Phil. Mag. 37, 498—515. (1919). (6)

F. RIESZ. Sur la formule d'inversion de Fourier. Acta litt. ac Sci. Univ. Hung. 3, 235—241. (1927). (9)

H. L. RIETZ. Mathematical statistics. Chicago, 1927. (10)

R. SCHMIDT 1. Über divergente Folgen und lineare Mittelbildungen. Math. Ztschr. 22, 89—152. (1925). (10)

—— 2. Über das Borelsche Summierungsverfahren. Schriften der Köningsberger gelehrten Gesellschaft, 1, 202—256. (1925). (10)

—— 3. Die trigonometrische Approximation für eine Klasse von verallgemeinerten fastperiodischen Funktionen. Math. Ann. 100, 334—356. (1928). (5)

I. SCHOENBERG 1. Über total monotone Folgen mit stetiger Belegungsfunktion. Mat. Ztschr. 30, 761—768. (1929). (8)

—— 2. Über die asymptotische Verteilung reeller Zahlen mod 1. Mat. Ztschr. 28, 177—200. (1928). (8)

A. SCHUSTER 1. On interference phenomena. Phil. Mag. 37, 509—545. (1894). (1)

—— 2. The periodogram of magnetic declination. Camb. Phil. Trans. 18, 108. (1899). (2)

—— 3. The periodogram and its optical analogy. Proc. Roy. Soc. 77, 136—140. (1906). (2)

—— 4. The theory of optics. London, 1904. (1)

—— 5. On lunar und solar periodicities of earthquakes. Proc. Roy. Soc. London 61, 455—465. (1897). (2)

—— 6. On hidden periodicities. Terrestrial Magnetism 3, 13. (1897). (2)

—— 7. The periodogram of magnetic declination. Trans. Camb. Phil. Soc. 18, 107—135. (1900). (2)

H. STEINHAUS. Les probabilités dénombrables et leur rapport à la théorie de la mésure. Fund. Math. 3, 286—310. (1923). (10)

W. STEPANOFF. Über einige Verallgemeinerungen der fastperiodischen Funktionen. Math. Ann. 90, 473—492. (1925). (5)

G. SZEGÖ. Zur Theorie der fastperiodischen Funktionen. Math. Ann. 96, 378—382. (1926). (5)

G. I. TAYLOR. Diffusion by continuous movements. Proc. Lond. Math. Soc. 20, 196—212. (1920). (6)

E. C. TITCHMARSH 1. A contribution to the theory of Fourier transforms. Proc. Lond. Math. Soc. 23, 279—289. (1924). (9)

—— 2. Recent advances in science mathematics. Science progress 91, 372—386. (1929). (5)

C. DE LA VALLÉE POUSSIN 1. Sur les fonctions presque périodiques de H. Bohr. Annales de la Societé Scentifique de Bruxelles, A, 47, 141. (1927). (5)

—— 2. Sur les fonctions presque périodiques de H. Bohr. Note complementaire et explicative. A.S.S. Bruxelles, A, 48, 56—57. (1928). (5)

T. VIJAYARAGHAVAN 1. A Tauberian theorem. Jour. Lond. Math. Soc. 1, 113—120. (1926). (10)

—— 2. A theorem concerning the summability of series by Borel's method. Proc. Lond. Math. Soc. 27, 316—326. (1928). (10)

V. VOLTERRA. Leçons sur les fonctions de lignes. Paris, 1913. (10)

J. D. WALSH. A generalization of the Fourier cosine series. Trans. Am. Math. Soc. 21, 101—116. (1920). (4)

H. WEYL 1. Integralgleichungen und fastperiodische Funktionen. Math. Ann. 97, 338—356. (1926). (5)

—— 2. Beweis des Fundamentalsatzes in der Theorie der fastperiodischen Funktionen. Berliner Sitzungsber. 1926, 211—214. (5)

—— 3. Quantenmechanik und Gruppentheorie. Ztschr. f. Physik 46, 1—46. (1927). (10)

N. WIENER 1. On the representation of functions by trigonometrical integrals. Math. Ztschr. 24, 575—617. (1925). (8)

—— 2. The harmonic analysis of irregular motion. J. Math. Phys. Mass. Inst. Technology 5, 99—122. (1925). (8)

—— 3. The harmonic analysis of irregular motion II. J.M.P.M.I.T. 5, 158—191. (1926). (8)

—— 4. The spectrum of an arbitrary function. Proc. Lond. Math. Soc. 27, 487—496. (1928). (8)

—— 5. Coherency matrices and quantum theory. J.M.P.M.I.T. 7, 109—125. (1928). (8)

—— 6. Harmonic analysis and the quantum theory. Jour. of Franklin Institute 207, 525—534. (1929). (8)

—— 7. The average of an analytic functional. Proc. Nat. Acad. Sci. 7, 253—260. (1921). (6)

—— 8. The average of an analytic functional and the Brownian motion. Proc. Nat. Acad. Sci. 7, 294—298. (1921). (6)

33—29764. *Acta mathematica.* 55. Imprimé le 9 mai 1930.

N. WIENER 9. The quadratic variation of a function and its Fourier coefficients. J.M.P.M.I.T. 3, 72—94. (1924). (8)

—— 10. Differential-space. J.M.P.M.I.T. 2, 131—174. (1923). (6)

—— 11. The average value of a functional. Proc. Lond. Math. Soc. 22, 454—467. (1922). (6)

—— 12. On a theorem of Bochner and Hardy. Jour. Lond. Math. Soc. 2, 118—123. (1927). (10)

—— 13. A new method in Tauberian theorems. J.M.P.M.I.T 7, 161—184. (1928). (10)

—— 14. The spectrum of an array and its application to the study of the translation properties of a simple class of arithmetical functions. J.M.P.M.I.T. 6, 145—157. (1927). (8)

—— 15. Harmonic analysis and group theory. J.M.P.M.I.T. 8, 148—154. (1929). (8)

—— 16. The operational calculus. Math. Ann. 95, 557—584. (1926). (8)

—— 17. Verallgemeinerte trigonometrische Entwicklungen. Gött. Nachrichten, 1925, 151—158. (8)

A. WINTNER. Spektraltheorie der unendlichen Matrizen. Leipzig, 1929. (5)

W. H. YOUNG. On non-harmonic Fourier series. Proc. Lond. Math. Soc. 18, 307—335. (1919). (4)

Footnotes.

1. In accordance with the listing adopted in the bibliography, the following references will give the background of the corresponding sections.

   1. PLANCHEREL 1, 2, 3; and in general, papers under rubric (9).
   2. Papers under rubric (2).
   3 and 4. WIENER 1, 2, 3, 4, 17.
   5. SCHMIDT 1, 2; VIJAYARAGHAVAN 1, 2; WIENER 13; JACOB 1.
   6. BOCHNER 7.
   7. HAHN 1. 2; DORN; JACOB 2, 3.
   8. BERRY 1, 2.
   9. POINCARÉ; WIENER 5, 6; WEYL 3; RIETZ.
   10. WIENER 15.
   11. MAHLER; WIENER 14.
   12. WIENER 14.
   13. WIENER 2, 3, 7, 8, 10, 11; EINSTEIN; TAYLOR; RAYLEIGH 1, 2, 3, 4, 5, 7, 8.
   14, 15, 16. Papers under rubric (5).
   17. MUCKENHOUPT.

2. SCHUSTER 5., p. 464.
3. HOBSON, § 492.
4. WEYL 3.
5. VOLTERRA, Ch. 7.
6. BOREL; STEINHAUS.

TAUBERIAN THEOREMS*

INTRODUCTION.

Numerous important branches of mathematics and physics concern themselves with the asymptotic behavior of functions for very large or very

* Received July 20, 1931.

Reprinted from *Ann. of Math.*, *33*, 1932, pp. 1–100. (Courtesy of the *Annals of Mathematics*.)

small values of their arguments, or of certain parameters. Statistical mechanics is that branch of mechanics which concerns itself, not with an individual dynamical system of a finite number of degrees of freedom, but with the asymptotic behavior of dynamical systems as the number of degrees of freedom increases without limit. The analytical theory of numbers is likewise concerned with the behavior of assemblages of whole numbers as the size of these assemblages increases. To this domain of ideas belong asymptotic series in analysis, and a whole order of concepts clustering about the operational calculus of Heaviside and the Fourier series and integrals.

The Fourier integral is of peculiar interest in the study of asymptotic problems. If $f(x)$ is a function of Lebesgue class L_2, by a theorem of Plancherel,[1] there is related to it another function $g(u)$, also of L_2, defined by the equation

$$g(u) = \underset{A\to\infty}{\text{l. i. m.}} \frac{1}{\sqrt{2\pi}} \int_{-A}^{A} f(x)\, e^{iux}\, dx \tag{0.01}$$

(where l. i. m. stands for "limit in the mean"), such that

$$\int_{-\infty}^{\infty} |g(u)|^2\, du = \int_{-\infty}^{\infty} |f(x)|^2\, dx, \tag{0.02}$$

and

$$f(x) = \underset{A\to\infty}{\text{l. i. m.}} \frac{1}{\sqrt{2\pi}} \int_{-A}^{A} g(u)\, e^{-iux}\, du. \tag{0.03}$$

The functions $f(x)$ and $g(u)$ have the reciprocal relation, that asymptotic properties of each correspond to local properties of the other. They are known as Fourier transforms of one another. It is easy to see that $f(x+\lambda)$ and $g(u)\,e^{-iu\lambda}$ are likewise Fourier transforms of one another.

There is a large class of asymptotic problems in which the asymptotic property to be investigated is connected in some obvious and simple way with the entire class of functions $f(x+\lambda)$. Since the Fourier transforms of the functions of this class only differ by factors $e^{-iu\lambda}$, independent of the particular function $f(x)$, Fourier transformation is here a peculiarly useful tool. An example of such a problem is the investigation of the asymptotic behavior of the integral

$$\int_{-\infty}^{\infty} k(\lambda)\, f(x+\lambda)\, d\lambda. \tag{0.04}$$

Many problems which on first investigation do not appear to be concerned with integrals of the above sort may be put into such a form by an elementary transformation of variables. For example, if we put

[1] Cf. Plancherel (1), Titchmarsh (3), Mellin (1), Wiener (7). Cf. Bibliography on p. 94.

(0.05) $x = \log \xi, \quad \lambda = \log \mu, \quad f(x) = \varphi(\xi), \quad \mu k(\lambda) = Q(\mu),$

the integral in question assumes the form

$$\int_0^\infty Q(\mu)\, \varphi(\xi\mu)\, d\mu. \tag{0.06}$$

Thus the study of the asymptotic properties of this integral also fall under those accessible through Fourier developments.

In 1925 a paper by Robert Schmidt[2] appeared in which the class of theorems known as Tauberian was brought into relation with the asymptotic properties of integrals of this type. Tauberian theorems gain their name from a theorem published by A. Tauber[3] in 1897, to the effect that if

$$\lim_{x \to 1-0} \sum_0^\infty a_n x^n = A, \tag{0.07}$$

and

$$a_n = o(1/n), \tag{0.08}$$

then

$$\sum_0^\infty a_n = A. \tag{0.09}$$

This is a conditioned converse of Abel's theorem, which stated that (0.07) follows from (0.09), without the mediation of any hypothesis such as (0.08). Such conditioned inverses of Abel's theorem, and of other analogous theorems which assert that the convergence of a series implies its summability by a certain method to the same sum, have been especially studied by G. H. Hardy and J. E. Littlewood, and have been termed by them *Tauberian.*

It is the service of Hardy and Littlewood[4] to have replaced hypothesis (0.08) by hypotheses of the form

$$a_n = O\left(\frac{1}{n}\right), \tag{0.10}$$

or even of the form

$$n a_n > -K. \tag{0.11}$$

The importance of these generalizations is scarcely to be exaggerated. They far exceed in significance Tauber's original theorem. The work of Hardy and Littlewood, unlike that of Tauber, makes very appreciable demands on analytical technique, and is capable, among other things, of supplying the gaps in Poisson's imperfect discussion of the convergence

[2] Schmidt (2).

[3] Tauber (1).

[4] Hardy (2), Littlewood (1), Hardy and Littlewood (4), (10), (13), (18) etc.

of the Fourier series. For these reasons, I feel that it would be far more appropriate to term these theorems Hardy-Littlewood theorems, were it not that usage has sanctioned the other appelation.

As we said, Tauberian theorems reduce to theorems on the conditioned equivalence of the asymptotic values of certain averages. The auxiliary conditions of Hardy and Littlewood, together with certain more extended conditions given by Schmidt[5] and discussed by Vijayaraghavan[6] and Szász,[7] become restrictions on the magnitude of the function or mass distribution of which the average is to be taken. An even further reduction eliminates this function from all consideration, and finds the essence of the Tauberian theorem in the linear properties of the weighting functions used in determining the average in question. The linear properties which are of importance concern the closure of the set of functions derivable from a given weighting function by translation, and the first chapter of this monograph is devoted to the study of such closure properties. The second chapter is occupied with the formulation and proof of the fundamental Tauberian theorems concerning averages, and the third with the transformation of these theorems into a recognized Tauberian form.

Some of the most interesting applications of Tauberian theorems have been in the analytic theory of numbers. Here two different avenues of approach are possible. Hardy and Littlewood[8] have shown that the prime number theorem, to the effect that the number of primes less than n is asymptotically $n/\log n$, is entirely equivalent to the Tauberian theorem concerning Lambert series, that if

$$\lim_{x\to 1-0} (1-x) \sum_0^\infty \frac{n\, a_n\, x^n}{1-x^n} = A \tag{0.12}$$

and

$$n\, a_n > -K, \tag{0.11}$$

then

$$\sum_0^\infty a_n = A. \tag{0.09}$$

This theorem falls under the category of theorems demonstrable by the methods of the present paper, and involves no further information concerning the Riemann zeta function $\zeta(\sigma+\tau i)$ than that this must be free of zeros on the line $\sigma = 1$. This theorem was already contained in the author's first note on his method, dated 1928, and establishes the usefulness of Lambert series in the proof of the prime number theorem, which had

[5] Schmidt (1) and (2).

[6] Vijayaraghavan (1) and (2).

[7] Szász (1) and (2).

[8] Hardy and Littlewood (11).

frequently been questioned[9]. The proof was however needlessly complicated by the fact that after a good deal of labor was spent in transforming the author's average theorem into recognizable Tauberian form, this form did not appear to be especially direct in its relation to the prime number theorem. The present memoir gives a much more direct proof, starting from a theorem of the average type.

The other approach to the prime number theorem employs a theorem of S. Ikehara[10], which is itself a generalization of a theorem of Landau. Landau[11] showed that if $\sum a_n n^{-x}$ is a Dirichlet series with positive coefficients, representing a function $\psi(x)$ analytic on and to the right of $\Re(x) = 1$, except for a pole of order 1 at $x = 1$, for which the residue is A, and if the function in question is $O(|x|^k)$ for $\Re(x) \geqq 1$, then

$$\lim_{n\to\infty} \frac{1}{n} \sum_0^n a_l = A. \tag{0.13}$$

Ikehara proved by a Tauberian theorem that the condition that $\psi(x) = O(|x|^k)$ is inessential. By applying the theorem to $\psi(x) = -\zeta'(x)/\zeta(x)$ it is at once seen that the prime number theorem again follows from the fact that the Riemann zeta function has no zeros on the line $\Re(x) = 1$.

Chapter V is devoted to miscellaneous applications of Tauberian theorems. Among these perhaps the most important are to trigonometric developments and their summability. Here a theorem of Ramanujan[12] on Fourier transforms is of service. A particular application of Tauberian theorems to the criterion of Young[13] for the convergence of a Fourier series is due to Dr. Littauer[14], and is here discussed. We also take up the application of Tauberian theorems to asymptotic series and Wintner's[15] work.

There is a further field of application for Tauberian theorems where we need to introduce something like Robert Schmidt's[16] notion of "gestrahlte Matrizen". This we discuss in Chapter VI. In this category we find certain theorems of Hardy and Littlewood[17] related to Abel summation, as also the Tauberian theorems that concern themselves with Borel summation[18].

[9] Wiener (4).

[10] Ikehara (1).

[11] Landau (3).

[12] Hardy and Littlewood (8).

[13] Young (1).

[14] Littauer (2).

[15] Wintner (1).

[16] Schmidt (2).

[17] Cf. Hardy and Littlewood (10), (13), (18); Karamata (1), (2), (3), (4); Szász (1), (2); Doetsch (3); etc.

[18] Cf. Hardy and Littlewood (2), (3); Doetsch (4); Schmidt (1); Vijayaraghavan (1); Wiener (4).

In Chapter VII, we discuss certain theorems not strictly of a Tauberian nature, in that they involve no positiveness or boundedness condition. We here follow an earlier paper[19] in which the author proved the Hardy-Littlewood necessary and sufficient condition for the summability of a Fourier series. There our result was not a "best possible" one, as the work of Bosanquet[20] and Paley has demonstrated. At present, our method has succeeded in yielding their strict theorem.

Chapter VIII is devoted to the development of certain Tauberian theorems intimately connected with our generalized harmonic analysis[21]. These suggest certain new definitions of summability, and we discuss definitions of this sort.

The genesis of the present paper may deserve a word of comment. In the preparation of a previous investigation on generalized harmonic analysis, the author found himself obliged to make use of certain theorems communicated to him by Mr. A. E. Ingham and of a Tauberian nature. A correspondence arose with Professor Hardy, finally resulting in a number of papers of a Tauberian character. While interested in theorems of this type, the author had the work of Dr. Robert Schmidt brought to his notice, and at Dr. Schmidt's suggestion, began to search for a method of combining the trigonometric attack of his own earlier papers with the generality of the Schmidt standpoint. It was only by a radical change in the manner in which trigonometric methods were applied that this attempt succeeded. This change consisted in a logarithmic change of base before the introduction of the harmonic analysis. It was quite along lines contemplated by Schmidt, who correlated with his Tauberian theorems a certain moment problem. Schmidt, however, found in the problem of uniqueness the significant aspect of this moment problem. The present author looked for it rather in the problem of the existence of a solution together with certain associated problems of approximation. It was Dr. Schmidt himself who furnished the experimentum crucis which established the greater scope of the methods of this paper. He suggested that the Tauberian theorem for Lambert series had resisted his methods of proof, and that it might be desirable to try on it the edge of any new method.

The origination of a new method is but the prelude to a large amount of detailed application before its power can be judged or its limitations defined. In this task the author has had the valuable aid of his students, Drs. S. B. Littauer and S. Ikehara. He also wishes to express his gratitude for the criticisms of Professors Hardy and Tamarkin at more than one stage of his work.

[19] Cf. Hardy and Littlewood (12); Wiener (5).

[20] Bosanquet (1); Paley (1).

[21] Wiener (7).

CHAPTER I.

THE CLOSURE OF THE SET OF TRANSLATIONS OF A GIVEN FUNCTION.

1. **Closure in class L_2.** Let $f(x)$ be a function, real or complex, defined for all real arguments over $(-\infty, \infty)$. Let it belong to Lebesgue class L_2—that is, let it be measurable, and let

$$\int_{-\infty}^{\infty} |f(x)|^2 \, dx \tag{1.01}$$

be finite. We shall term the class of all functions $f(x+\lambda)$ for all real values of λ the class of *translations of* $f(x)$. We wish to know when this class is closed or complete—that is, when it is possible, whenever a function $F(x)$ from L_2 is given, and any positive quantity ε, to find a function $F_1(x)$ of the form

$$F_1(x) = \sum_1^n A_k f(x+\lambda_k), \tag{1.02}$$

such that

$$\int_{-\infty}^{\infty} |F(x) - F_1(x)|^2 \, dx \leqq \varepsilon. \tag{1.03}$$

The situation is governed by the following theorem:

THEOREM I. *The necessary and sufficient condition for the set of all translations of $f(x)$ to be closed is that the real zeros of its Fourier transform*

$$g(u) = \underset{A\to\infty}{\text{l. i. m.}} \frac{1}{\sqrt{2\pi}} \int_{-A}^{A} f(x)\, e^{iux} \, dx \tag{0.01}$$

should form a set of zero measure.

The existence of this Fourier transform results from a familiar theorem of Plancherel.[1] This theorem further requires that

$$\int_{-\infty}^{\infty} |g(u)|^2 \, du = \int_{-\infty}^{\infty} |f(x)|^2 \, dx \tag{0.02}$$

and that

$$f(x) = \underset{A\to\infty}{\text{l. i. m.}} \frac{1}{\sqrt{2\pi}} \int_{-A}^{A} g(u)\, e^{-iux} \, du. \tag{0.03}$$

The Fourier transform of $f(x+\lambda)$ is $e^{-iu\lambda} g(u)$.

On account of (0.02), properties of convergence in the mean are conserved under a Fourier transformation, for these concern merely the integral of the square of the modulus of the difference of two functions, which is equal to the integral of the square of the difference of their Fourier transforms. Accordingly, Theorem I is a corollary of:

Lemma Ia. *If $g(u)$ is a function belonging to L_2, then the set of functions $e^{-iu\lambda}g(u)$ is closed when and only when the zeros of $g(u)$ form a set of zero measure.*

Since every zero of $g(u)$ is also a zero of $e^{-iu\lambda}g(u)$, the necessity of this condition is at once obvious, for if

$$G_1(u) = \sum_1^n A_k e^{-iu\lambda_k} g(u) \tag{1.04}$$

is any linear combination of the functions $e^{-iu\lambda_k}g(u)$, then

$$\int_A^B |G_1(u) - 1|^2 du \tag{1.05}$$

must equal or exceed the measure of the set of the points between A and B where $g(u)$ vanishes. That is, if the measure of this set is not zero, the function which is 1 between A and B and zero elsewhere cannot be a limit in the mean of functions $G_1(u)$.

Now as to the sufficiency of the condition in Lemma Ia: let

$$\varphi(u) = \begin{cases} 1 - \dfrac{|u|}{A}, & [|u| \leqq A]; \\ 0, & [|u| > A]. \end{cases} \tag{1.06}$$

Then

$$\varphi(u) = \frac{1}{\pi A} \int_{-\infty}^{\infty} e^{-iu\mu} \frac{1 - \cos A\mu}{\mu^2} d\mu. \tag{1.07}$$

This is an absolutely convergent integral, and hence a limit of a sequence of polynomials

$$\sum A_n e^{-iu\lambda_n} \tag{1.08}$$

converging boundedly, and uniformly over every finite interval. Hence

$$e^{-i\frac{un\pi}{A}} \varphi(u) g(u) \tag{1.09}$$

is a limit in the mean of such functions as those in (1.04).

Now let $\psi(u)$ be a quadratically summable function satisfying the conditions

$$\int_{-A}^{A} \psi(u)\varphi(u)g(u) e^{\frac{-iun\pi}{A}} du = 0. \tag{1.10}$$

$$[n = \cdots, -2, -1, 0, 1, 2, \cdots]$$

Since $\psi(u)\varphi(u)g(u)$ is absolutely summable,[22] we shall have

$$\psi(u)\varphi(u)g(u) = 0 \tag{1.11}$$

[22] This follows from the Parseval theorem for Fourier integrals.

except over a set of measure zero.[23] If $g(u)$ does not vanish except over a set of measure zero, since $\varphi(u)$ has no zeros, $\psi(u)$ vanishes over $(-A, A)$ except over a set of measure zero. Thus if $g(u) \neq 0$, there is no quadratically summable function not almost everywhere zero, orthogonal to every function $e^{\frac{-iun\pi}{A}} \varphi(u) g(u)$ over $(-A, A)$. This means that every function of L_2 vanishing everywhere outside $(-A, A)$ is a limit in the mean of polynomials[24] in $e^{\frac{-iun\pi}{A}} \varphi(u) g(u)$ and hence in $e^{-iu\lambda_n} g(u)$. Every function of L_2 is however the limit in the mean of a sequence of functions of L_2 vanishing for large arguments. Thus Lemma Ia and Theorem I are established.

2. **Closure in class L_1.** The Lebesgue class L_1 consists of all measurable functions $f(x)$ which are absolutely integrable over $(-\infty, \infty)$. If $f(x)$ is such a function, its Fourier transform

$$g(u) = \frac{1}{\sqrt{2\pi}} \int_{-\infty}^{\infty} f(x) e^{iux} dx \tag{2.01}$$

of course exists, and is bounded and continuous. As before, the Fourier transform of $f(x+\lambda)$ is $e^{-iu\lambda} g(u)$.

We shall say that a class C of functions $\varphi_\lambda(x)$ of L_1 is closed L_1 if, whenever $F(x)$ is a function of L_1 and ε is a positive number, there is a polynomial

$$F_1(x) = \sum_1^n A_k \varphi_{\lambda_k}(x) \tag{2.02}$$

of functions of C, such that

$$\int_{-\infty}^{\infty} | F(x) - F_1(x) | \, dx \leq \varepsilon. \tag{2.03}$$

We shall prove the following theorem:

Theorem II. *If $f(x)$ is a function of L_1, a necessary and sufficient condition for the set of all translations of $f(x)$ to be closed L_1 is that its Fourier transform*

$$g(u) = \frac{1}{\sqrt{2\pi}} \int_{-\infty}^{\infty} f(x) e^{iux} dx \tag{2.01}$$

should have no real zeros.

[23] This follows from the Riemann theorem on the unicity of the function with a given Fourier series.

[24] Here we make use of the familiar theorem that if any linear class of functions of L_2 is given, every function of L_2 is the sum of a function expressible in the mean by the functions of the class and a function orthogonal to all functions of the class.

The necessity of this condition is again obvious. If $g(u_1) = 0$, the same is true for the argument u_1 of the Fourier transform of

$$(1.02)\qquad F_1(x) = \sum_1^n A_k f(x+\lambda_k).$$

Let $f_1(x)$ be a function of L_1 with a Fourier transform $g_1(x)$ for which

$$g_1(u) \neq 0.$$

Then

$$(2.04)\qquad \int_{-\infty}^{\infty} |F_1(x)-f_1(x)|\; dx \geqq \left|\int_{-\infty}^{\infty} (f_1(x)-F_1(x))\, e^{iu_1x}\, dx\right| = \sqrt{2\pi}\; |g_1(u_1)| > 0.$$

This proves that f_1 is not a limit in the mean (L_1) of any sequence of F_1's.

To prove the sufficiency of the condition of Theorem II, we shall have recourse to a sequence of lemmas.

Lemma IIa. *If*

$$(2.05)\qquad f_1(x) = \sum_{-\infty}^{\infty} a_n e^{inx}, \qquad f_2(x) = \sum_{-\infty}^{\infty} b_n e^{inx},$$

and

$$(2.06)\qquad \sum_{-\infty}^{\infty} |a_n| = A < \infty, \qquad \sum_{-\infty}^{\infty} |b_n| = B < \infty,$$

and if

$$(2.07)\qquad f_1(x)\, f_2(x) \sim \sum_{-\infty}^{\infty} c_n e^{inx}$$

then

$$(2.08)\qquad \sum_{-\infty}^{\infty} |c_n| \leqq AB.$$

The proof is immediate, for

$$(2.09)\qquad \sum_{-\infty}^{\infty} |c_n| = \sum_{n=-\infty}^{\infty} \left|\sum_{k=-\infty}^{\infty} a_k b_{n-k}\right| \leqq \sum_{n=-\infty}^{\infty} \sum_{k=-\infty}^{\infty} |a_k|\; |b_{n-k}| = \sum_{-\infty}^{\infty} |a_n| \sum_{-\infty}^{\infty} |b_n| = AB.$$

Lemma IIb. *If $f(x)$ is a function of period 2π, and if at every point y, there is a function $f_y(x)$, coincident with $f(x)$ over some interval $(y-\varepsilon_y, y+\varepsilon_y)$, of period 2π, and such that its Fourier series converges absolutely, then the Fourier series of $f(x)$ converges absolutely*[25].

[25] We shall say that a Fourier series converges absolutely when the sum of the moduli of the coefficients converges, the series being represented in complex exponential form. This is of course equivalent to the absolute convergence of the series in this form for any single real argument.

To prove this, let us reflect that by the Heine-Borel theorem, any period of $f(x)$ may be covered by a finite number of overlapping intervals $(y-\varepsilon_y, y+\varepsilon_y)$. Let such an interval be (A, B), let its neighbor to the left have its right-hand end-point at C, and let its neighbor to the right have its left-hand end-point at D. There will be no loss in generality if we suppose that $A<C<D<B$. Let us define $\varphi_y(x)$ by

$$(2.10) \qquad \varphi_y(x) = \begin{cases} 0\ ; & [-\pi \leqq x < A] \\ \dfrac{x-A}{C-A}; & [\ A \leqq x < C] \\ 1\ ; & [\ C \leqq x < D] \\ \dfrac{B-x}{B-D}; & [\ D \leqq x < B] \\ 0 & [\ B \leqq x < \pi]. \end{cases}$$

It is easy to show that the Fourier series of $\varphi_y(x)$ converges absolutely. Thus the Fourier series of

$$(2.11) \qquad \varphi_y(x)\, f(x) = \varphi_y(x)\, f_y(x)$$

converges absolutely, by Lemma IIa. However, if we add together the functions $\varphi_y(x)$ for a set of intervals completely covering a period of $f(x)$, the sum will be $f(x)$ itself. Thus $f(x)$ will be the sum of a finite number of functions with absolutely convergent Fourier series and must itself have an absolutely convergent Fourier series.

LEMMA IIc. *If*

$$(2.12) \qquad f(x) = \sum_{-\infty}^{\infty} a_n e^{inx}$$

and

$$(2.13) \qquad |a_0| > \sum_{-\infty}^{-1} |a_n| + \sum_{1}^{\infty} |a_n|,$$

then the Fourier series of $1/f(x)$ *converges absolutely.*

We may write

$$(2.14) \qquad \begin{aligned} \frac{1}{f(x)} &= \frac{1}{a_0 + \sum_{-\infty}^{-1} a_n e^{inx} + \sum_{1}^{\infty} a_n e^{inx}} \\ &= \frac{1}{a_0}\left\{1 - \frac{1}{a_0}\left[\sum_{-\infty}^{-1} a_n e^{inx} + \sum_{1}^{\infty} a_n e^{inx}\right]\right. \\ &\qquad \left. + \frac{1}{a_0^2}\left[\sum_{-\infty}^{-1} a_n e^{inx} + \sum_{1}^{\infty} a_n e^{inx}\right]^2 - \cdots\right\}. \end{aligned}$$

That this is more than formally true, and that $1/f(x)$ is continuous, results from the fact that the geometrical progression

$$(2.15)\quad \frac{1}{|a_0|}\left\{1+\frac{1}{|a_0|}\left[\sum_{-\infty}^{-1}|a_n|+\sum_{1}^{\infty}|a_n|\right]+\frac{1}{a_0^2}\left[\sum_{-\infty}^{-1}|a_n|+\sum_{1}^{\infty}|a_n|\right]^2+\cdots\right\}$$

converges. Term by term, series (2.15) is greater than or equal to the sum of the sums of the moduli of the coefficients of the Fourier series of the successive terms of (2.14), and hence is greater than or equal to the sum of the moduli of the coefficients of the Fourier series of $1/f(x)$.

LEMMA IId. *Let*

$$(2.16)\qquad \sum_{-\infty}^{\infty}|a_n| = A<\infty,$$

and let

$$(2.17)\qquad f(x) = \sum_{-\infty}^{\infty} a_n e^{inx}.$$

Let

$$f(y) \neq 0.$$

Then there is a neighborhood $(y-\varepsilon,\ y+\varepsilon)$ *of* y *and a function* $f_y(x)$ *with absolutely convergent Fourier series such that over* $(y-\varepsilon,\ y+\varepsilon)$,

$$(2.18)\qquad f_y(x) = f(x)$$

and that the Fourier series of $f_y(x)$, *let us say*

$$(2.19)\qquad \sum_{-\infty}^{\infty} c_n e^{inx}$$

has the property that

$$(2.20)\qquad |c_0| > \sum_{-\infty}^{-1}|c_n| + \sum_{1}^{\infty}|c_n|.$$

There is manifestly no restriction in taking y to be 0. Let us introduce the auxiliary function

$$(2.201)\qquad \varphi_\varepsilon(x) = \begin{cases} 1 & ;\ [|x|<\varepsilon] \\ 2-\dfrac{|x|}{\varepsilon}; & [\varepsilon\leqq|x|<2\varepsilon] \\ 0 & .\ [2\varepsilon\leqq|x|]. \end{cases}$$

This will have the Fourier series

$$(2.202)\qquad \frac{3\varepsilon}{2\pi}+\sum_{1}^{\infty}\frac{e^{inx}+e^{-inx}}{\pi n^2 \varepsilon}(\cos\varepsilon n-\cos 2\varepsilon n).$$

If we represent the function $\{f(x)\,\varphi_\varepsilon(x)+f(0)\,[1-\varphi_\varepsilon(x)]\}$ by the Fourier series

$$\sum_{-\infty}^{\infty} c_n e^{inx},$$

we shall have

$$c_n = \frac{3\varepsilon}{2\pi} a_n + \sum_{k=1}^{\infty} \frac{a_{n-k} + a_{n+k}}{\pi k^2 \varepsilon} (\cos \varepsilon k - \cos 2\varepsilon k)$$

$$- \sum_{-\infty}^{\infty} a_k \frac{\cos \varepsilon n - \cos 2\varepsilon n}{\pi n^2 \varepsilon}; \qquad [n \neq 0],$$

(2.203)

$$c_0 = a_0 + \sum_{1}^{\infty} (a_{-k} + a_k) \left[1 + \frac{\cos \varepsilon k - \cos 2\varepsilon k}{\pi k^2 \varepsilon} - \frac{3\varepsilon}{2\pi} \right].$$

Hence

$$\lim_{\varepsilon \to 0} |c_0| = \left| \sum_{-\infty}^{\infty} a_k \right| = |f(0)| > 0, \tag{2.204}$$

and

$$\sum_{1}^{\infty} [|c_n| + |c_{-n}|]$$

$$\leqq \sum_{m=-\infty}^{\infty} |a_m| \left\{ \sum_{n=-\infty}^{\infty}{}'' \left| \frac{\cos \varepsilon n - \cos 2\varepsilon n}{\pi n^2 \varepsilon} - \frac{\cos \varepsilon (m-n) - \cos 2\varepsilon (m-n)}{\pi (m-n)^2 \varepsilon} \right| \right.$$

(2.205)

$$\left. + \left| \frac{\cos \varepsilon m - \cos 2\varepsilon m}{\pi m^2 \varepsilon} - \frac{3\varepsilon}{2\pi} \right| \right\}$$

$$= \sum_{-\infty}^{\infty} |a_m| A_m.$$

Here $\sum''$ means that the values 0 and m are omitted.

We have

$$\left| \frac{\cos \varepsilon n - \cos 2\varepsilon n)}{\pi n^2 \varepsilon} \right| = \left| \frac{2 \sin \frac{\varepsilon n}{2} \sin \frac{3\varepsilon n}{2}}{\pi n^2 \varepsilon} \right| \leqq \begin{cases} \frac{3\varepsilon}{2\pi}; \\ \frac{2}{\pi n^2 \varepsilon}; \end{cases} \tag{2.206}$$

and

$$\left| \frac{\cos \varepsilon n - \cos 2\varepsilon n}{\pi n^2 \varepsilon} - \frac{\cos \varepsilon (m-n) - \cos 2\varepsilon (m-n)}{\pi (m-n)^2 \varepsilon} \right|$$

$$= \left| \int_{n-m}^{n} \frac{d}{dx} \left(\frac{\cos \varepsilon x - \cos 2\varepsilon x}{\pi x^2 \varepsilon} \right) dx \right| \tag{2.21}$$

$$\leqq \frac{\varepsilon^2}{\pi} \left| \int_{n-m}^{n} \left| \max_{y} \frac{d}{dy} \frac{\cos y - \cos 2y}{y^2} \right| dx \right| \leqq |m| \varepsilon^2 c.$$

This formula may be applied in the limiting form whenever $m = n$ or $m = 0$. It results from (2.205) and (2.206) that for all sufficiently small ε

$$A_m \leqq 4 \left[\frac{1}{\varepsilon} + 1 \right] \frac{3\varepsilon}{2\pi} + \frac{8}{\pi\varepsilon} \sum_{[1/\varepsilon]}^{\infty} n^{-2} < \frac{15}{\pi}, \tag{2.22}$$

and from (2.205) and (2.21) that for all sufficiently small ε

$$A_m \leqq 2[\varepsilon^{-3/2}+1]\,|\,m\,|\,\varepsilon^2 c + \frac{8}{\pi\varepsilon}\sum_{[\varepsilon^{-3/2}]}^{\infty} n^{-2} \tag{2.23}$$
$$< \varepsilon^{1/2}[2\,|\,m\,|\,c+9/\pi].$$

Combining (2.205), (2.22) and (2.23), we get

$$\lim_{\varepsilon\to 0}\sum_{-\infty}^{\infty}|\,a_m\,|\,A_m = 0, \tag{2.24}$$

which with (2.204) yields us

$$\sum_{1}^{\infty}[|\,c_n\,|+|\,c_{-n}\,|] < |\,c_0\,|. \tag{2.25}$$

If we take

$$f_y(x) = f(x)\,\varphi_\varepsilon(x) + f(0)\,(1-\varphi_\varepsilon(x)), \tag{2.26}$$

Lemma IId is established.

As a direct consequence of Lemmas IIb, IIc, and IId we obtain:

LEMMA IIe. *If $f(x)$ is a function with an absolutely convergent Fourier series, which nowhere vanishes for real arguments, $1/f(x)$ has an absolutely convergent Fourier series.*

To prove this let us note that, in every neighborhood, by IId, $f(x)$ coincides with a function which by IIc has a reciprocal with an absolutely convergent Fourier series. By IIa, since in every neighborhood $1/f(x)$ coincides with a function with an absolutely convergent Fourier series, its Fourier series converges absolutely.

LEMMA IIf. *Let $f(x)$ be a function with an absolutely convergent Fourier series over the interval $(-\pi, \pi)$, and vanishing over neighborhoods including π and $-\pi$. Let $f(x)$ vanish everywhere outside $(-\pi, \pi)$. Let*

$$g(u) = \frac{1}{\sqrt{2\pi}}\int_{-\pi}^{\pi} f(x)\,e^{iux}\,dx. \tag{2.27}$$

Then the integral

$$\int_{-\infty}^{\infty}|\,g(u)\,|\,du \tag{2.28}$$

will converge. Conversely, if $f(x)$ vanishes over the region indicated, and (2.28) converges, its Fourier series will converge absolutely.

Let ε be such that $f(x)$ has no non-zero, values outside $(-\pi+2\varepsilon, \pi-2\varepsilon)$. Let us define $\varphi(x)$ by:

(2.29) $$\varphi(x) = \begin{cases} 1 & ; \ [|x| < \pi - \varepsilon] \\ \dfrac{\pi - |x|}{\varepsilon}; & [\pi - \varepsilon \leqq |x| < \pi] \\ 0 & [\pi \leqq |x|] \end{cases}$$

Let

(2.17) $$f(x) = \sum_{-\infty}^{\infty} a_n e^{inx}.$$

We shall have

(2.30) $$\varphi(x) = \frac{1}{\pi} \int_{-\infty}^{\infty} e^{-iux} \frac{\cos u(\pi - \varepsilon) - \cos u\pi}{u^2 \varepsilon} du,$$

and hence

(2.31) $$\begin{aligned} f(x) = f(x)\varphi(x) &= \frac{1}{\pi} \int_{-\infty}^{\infty} du \sum_{-\infty}^{\infty} a_n e^{-i(u-n)x} \frac{\cos u(\pi - \varepsilon) - \cos u\pi}{u^2 \varepsilon} \\ &= \frac{1}{\pi} \int_{-\infty}^{\infty} e^{iux} \left[\sum_{-\infty}^{\infty} a_n \frac{\cos(u+n)(\pi - \varepsilon) - \cos(u+n)\pi}{(u+n)^2 \varepsilon} \right] du, \end{aligned}$$

where the rearrangement is possible since both integral and sum form part of an absolutely convergent double summation process. Then

(2.32) $$g(u) = \sqrt{\frac{2}{\pi}} \sum_{-\infty}^{\infty} a_n \frac{\cos(u+n)(\pi - \varepsilon) - \cos(u+n)\pi}{(u+n)^2 \varepsilon}$$

will satisfy the condition that (2.28) shall be finite, and equation (2.27).

To prove the converse part of Lemma IIf, let us notice that as a result of the convergence of (2.28),

(2.33) $$\sum_{-\infty}^{\infty} \left| \int_{n-1/2}^{n+1/2} g(u)\, du \right|$$

converges. Now,

(2.34) $$\begin{aligned} \int_{n-1/2}^{n+1/2} g(u)\, du &= \frac{1}{\sqrt{2\pi}} \int_{-\pi}^{\pi} f(x)\, dx \int_{n-1/2}^{n+1/2} e^{iux}\, du \\ &= \frac{1}{\sqrt{2\pi}} \int_{-\pi}^{\pi} f(x) \frac{2 \sin x/2}{x} e^{inx}\, dx. \end{aligned}$$

Thus the Fourier series of

(2.35) $$f(x) \frac{2 \sin x/2}{x}$$

converges absolutely. Since the same is true of the Fourier series of

(2.36) $$\frac{x}{2 \sin x/2} \varphi(x)$$

as may be determined by a simple computation, it follows from Lemma IIa that the same is true for the Fourier series of $f(x)$.

LEMMA IIg. *Let $f(x)$ be a function of class L_1. Then*

$$\lim_{\varepsilon\to 0}\int_{-\infty}^{\infty}|f(x+\varepsilon)-f(x)|\,dx = 0. \tag{2.37}$$

This is a well-known theorem and is proved in Hobson's *Theory of Functions of a Real Variable.*

LEMMA IIh. *Let $f(x)$ be a function of class L_1. Then if $\Phi(y)$ is any function of class L_1, differing from 0 only over $(-\varepsilon,\varepsilon)$,*

$$\begin{aligned}\int_{-\infty}^{\infty}\left|f(x)\int_{-\infty}^{\infty}\Phi(y)\,dy-\int_{-\infty}^{\infty}f(x+y)\,\Phi(y)\,dy\right|dx\\ \leq\int_{-\infty}^{\infty}|\Phi(y)|\,dy\max_{|z|\leq\varepsilon}\int_{-\infty}^{\infty}|f(x+z)-f(x)|\,dx.\end{aligned} \tag{2.38}$$

The proof of this lemma merely depends on the inversion of the order of integration in an absolutely convergent double integral.

LEMMA IIi. *Let $f(x)$ be a function of class L_1. Then*

$$\lim_{n\to\infty}\int_{-\infty}^{\infty}\left|f(x)-\frac{1}{\pi n}\int_{-\infty}^{\infty}f(x+y)\frac{\sin^2 ny}{y^2}\,dy\right|dx = 0. \tag{2.39}$$

This is a theorem of the Fejér type. It depends on the splitting of

$$\Phi_0(y) = \frac{\sin^2 ny}{y^2}$$

into the sum $\Phi_1(y)+\Phi_2(y)$ where

$$\Phi_1(y) = \begin{cases}\Phi_0(y)\,(1-|y|\,n^{1/2}); & [|y|\leq n^{-1/2}]\\ 0; & [|y|>n^{-1/2}].\end{cases} \tag{2.40}$$

Lemmas IIg and IIh enable us to show that

$$\lim_{n\to\infty}\int_{-\infty}^{\infty}\left|f(x)-\frac{1}{\pi n}\int_{-\infty}^{\infty}f(x+y)\,\Phi_1(y)\,dy\right|dx = 0 \tag{2.41}$$

while

$$\begin{aligned}\lim_{n\to\infty}\int_{-\infty}^{\infty}\left|\frac{1}{\pi n}\int_{-\infty}^{\infty}f(x+y)\,\Phi_2(y)\,dy\right|dx\\ \leq\int_{-\infty}^{\infty}\left|f(x)\right|dx\lim_{n\to\infty}\frac{1}{\pi n}\int_{-\infty}^{\infty}\Phi_2(y)\,dy.\end{aligned} \tag{2.42}$$

Now

$$\frac{1}{\pi n}\int_{-\infty}^{\infty} \Phi_2(y)\,dy$$

$$= \frac{1}{\pi n}\int_{-\infty}^{\infty} \frac{\sin^2 ny}{y^2}\,dy - \frac{1}{\pi n}\int_{-n^{-1/2}}^{n^{-1/2}} (1-|y|n^{1/2})\frac{\sin^2 ny}{y^2}\,dy$$

$$(2.421)\qquad = \frac{1}{\pi}\int_{-\infty}^{\infty} \frac{\sin^2 z}{z^2}\,dz - \frac{1}{\pi}\int_{-n^{1/2}}^{n^{1/2}} (1-|z|n^{-1/2})\frac{\sin^2 z}{z^2}\,dz$$

$$= \frac{2}{\pi}\int_{n^{1/2}}^{\infty} \frac{\sin^2 z}{z^2}\,dz + \frac{2}{\pi n^{1/2}}\int_{0}^{n^{1/2}} \frac{\sin^2 z}{z}\,dz$$

$$= O(n^{-1/2}\log n).$$

Hence

$$(2.422)\qquad \lim_{n\to\infty}\int_{-\infty}^{\infty}\left|\frac{1}{\pi n}\int_{-\infty}^{\infty} f(x+y)\,\Phi_2(y)\,dy\right|dx = 0,$$

and Lemma IIi is established.

We are now in a position to proceed to the proof of Theorem II. Let $F(x)$ be any function of L_1. By Lemma IIi, we can find a function $F_\eta(x)$ of the form

$$(2.43)\qquad \frac{1}{\pi n}\int_{-\infty}^{\infty} F(x+y)\,\frac{\sin^2 ny}{y^2}\,dy,$$

such that

$$(2.44)\qquad \int_{-\infty}^{\infty} |F(x)-F_\eta(x)|\,dx<\eta.$$

The Fourier transform $H_1(u)$ of this function $F_\eta(x)$ is

$$\frac{1}{\sqrt{2\pi}}\int_{-\infty}^{\infty} e^{iux}\,dx\,\frac{1}{\pi n}\int_{-\infty}^{\infty} F(x+y)\,\frac{\sin^2 ny}{y^2}\,dy$$

$$(2.45)\qquad = \frac{1}{\sqrt{2\pi}}\int_{-\infty}^{\infty} F(x)\,e^{iux}\,dx\,\frac{1}{\pi n}\int_{-\infty}^{\infty} \frac{\sin^2 ny}{y^2}\,e^{-iuy}\,dy$$

$$= \begin{cases} \left(1-\dfrac{|u|}{2n}\right)\dfrac{1}{\sqrt{2\pi}}\displaystyle\int_{-\infty}^{\infty} F(x)\,e^{iux}\,dx; & [|u|<2n] \\ 0; & [|u|\geqq 2n] \end{cases}$$

and will vanish for all values of its argument u larger than $2n$ in modulus. The same will be the case with the Fourier transform $H_2(u)$ of

$$(2.46)\qquad \frac{1}{2\pi n}\int_{-\infty}^{\infty} f(x+y)\,\frac{\sin^2 2ny}{y^2}\,dy$$

which will vanish for values of its argument greater in modulus than $4n$, and only for such values. If we expand these Fourier transforms in Fourier series over $(-8n, 8n)$, by Lemma IIf, these series converge absolutely.

By an argument substantially identical with that used in proving Lemma IIb, we see that in the neighborhood of every point of $(-3n, 3n)$, there is a function coinciding locally with $1/H_2(u)$, and with an absolutely convergent Fourier series. It follows from Lemma IIa that the same is true of $H_1(u)/H_2(u)$ over the same interval. Since 0 enjoys this property also, the function which is $H_1(u)/H_2(u)$ over $(-3n, 3n)$ and 0 over the rest of a period $(-8n, 8n)$ has an absolutely convergent Fourier series, and hence, by Lemma IIf, an absolutely convergent Fourier integral. That is, we may write

$$H_1(u) = H_2(u)\, H_3(u) \tag{2.47}$$

where $H_3(u)$ is of the form

$$H_3(u) = \int_{-\infty}^{\infty} \Phi(x)\, e^{iux}\, dx \tag{2.48}$$

and

$$\int_{-\infty}^{\infty} |\Phi(x)|\, dx \tag{2.49}$$

converges. We shall then have

$$\begin{aligned}
&\frac{1}{\sqrt{2\pi}} \int_{-\infty}^{\infty} F_\eta(x)\, e^{iux}\, dx \\
&= \frac{1}{\sqrt{2\pi}} \int_{-\infty}^{\infty} \frac{e^{iux}}{2\pi n}\, dx \int_{-\infty}^{\infty} f(x+y)\, \frac{\sin^2 2ny}{y^2}\, dy \int_{-\infty}^{\infty} \Phi(z)\, e^{iuz}\, dz \\
&= \frac{1}{\sqrt{2\pi}} \int_{-\infty}^{\infty} e^{iux}\, dx \\
&\quad \times \left[\frac{1}{2\pi n} \int_{-\infty}^{\infty} f(x+\omega)\, d\omega \int_{-\infty}^{\infty} \frac{\sin^2 2ny}{y^2}\, \Phi(y-\omega)\, dy \right],
\end{aligned} \tag{2.50}$$

or

$$\begin{aligned}
&\int_{-\infty}^{\infty} e^{iux}\, dx \\
&\left[F_\eta(x) - \frac{1}{2\pi n} \int_{-\infty}^{\infty} f(x+\omega)\, d\omega \int_{-\infty}^{\infty} \frac{\sin^2 2ny}{y^2}\, \Phi(y-\omega)\, dy \right] = 0.
\end{aligned} \tag{2.51}$$

Here we may replace e^{iux} by any function with an absolutely convergent Fourier integral, and hence by functions positive over an arbitrarily small region and zero everywhere else. This can only be if

$$
\begin{aligned}
F_\eta(x) &= \frac{1}{2\pi n}\int_{-\infty}^{\infty} f(x+\omega)\,d\omega \int_{-\infty}^{\infty} \frac{\sin^2 2ny}{y^2}\,\Phi(y-\omega)\,dy \\
&= \int_{-\infty}^{\infty} f(x+\omega)\,\Psi(\omega)\,d(\omega).
\end{aligned} \tag{2.52}
$$

Here $\Psi(\omega)$ is absolutely integrable.

By Lemma IIh and Lemma IIg,

$$
\begin{aligned}
\lim_{n\to\infty} \int_{-\infty}^{\infty} \left| \int_{-\infty}^{\infty} f(x+\omega)\,\Psi(\omega)\,d\omega - \sum_{k=-n^2}^{n^2-1} f\left(x+\frac{k}{n}\right) \int_{\frac{k}{n}}^{\frac{k+1}{n}} \Psi(\omega)\,d\omega \right| dx \\
= 0.
\end{aligned} \tag{2.53}
$$

Combining this with (2.52) and Lemma IIi, we obtain Theorem II.

With only a slight modification of detail, we have proved:

THEOREM III. *Let $f_1(x)$ and $f_2(x)$ be two functions of L_1. Let*

$$
g_1(u) = \frac{1}{\sqrt{2\pi}} \int_{-\infty}^{\infty} f_1(x)\,e^{iux}\,dx \tag{2.54}
$$

and

$$
g_2(u) = \frac{1}{\sqrt{2\pi}} \int_{-\infty}^{\infty} f_2(x)\,e^{iux}\,dx. \tag{2.55}
$$

Let the set of points where $g_2(u) \neq 0$ consist only of inner points of the set where $g_1(u) \neq 0$. Then if $\varepsilon > 0$ there is a polynomial

$$
f_3(x) = \sum_1^N a_n f_1(x+\lambda_n) \tag{2.56}
$$

such that

$$
\int_{-\infty}^{\infty} |f_2(x) - f_3(x)|\,dx < \varepsilon. \tag{2.57}
$$

On the basis of Theorem III, we are in a position to prove the general:

THEOREM IV. *Let Σ be a class of functions of L_1. Then a necessary and sufficient condition for the class Σ_1, containing all functions $f(x+\lambda)$ whenever $f(x)$ belongs to Σ, to be closed L_1, is that there should be no real zero common to all the Fourier transforms of functions of Σ.*

The proof of the necessity of the condition of Theorem IV does not differ from the similar proof in the case of Theorem II. As to the sufficiency, it follows at once from Theorem III and the fact that the Fourier transform of a function of L_1 differs from 0 on an open set, that for every u, an ε may be assigned, such that

$$
e^{iux}\frac{\sin^2 \varepsilon x}{x^2} \tag{2.58}
$$

is a function approximable L_1 by polynomials in functions of Σ_1. Let $(-U, U)$ be any interval of frequencies. By the Heine-Borel theorem, this may be overlaid with a finite number of overlapping intervals $U_n-\varepsilon_n$, $U_n+\varepsilon_n$. Let us form

$$\sum e^{iU_n x}\frac{\sin^2 \varepsilon_n x}{x^2} \tag{2.59}$$

for this set of intervals. This will be a function of L_1, approximable L_1 by polynomials in the function of Σ_1, whose Fourier transform may be shown to be

$$\sum \frac{\varepsilon_n}{4}\left(1-2\left|\frac{U_n+u}{\varepsilon_n}\right|\right)\left[\operatorname{sgn}\left(1-2\left|\frac{U_n+u}{\varepsilon_n}\right|\right)+1\right] \tag{2.60}$$

and to have no zeros over $(-U, U)$. Thus by Theorem III, every function of L_1 with a Fourier transform vanishing outside $(-U+\varepsilon_n, U-\varepsilon_n)$ will be approximable L_1 by polynomials in the functions of Σ_1. At this stage, the introduction of Lemma IIi serves to complete the proof of Theorem IV.

In Theorems II and IV, the necessary part is nearly trivial and all the difficulty resides in the proof of sufficiency. There are certain applications, however, where it is precisely the necessary part of the theorem that is significant. We may prove a function to have no zeros if it is the Fourier transform of a function of L_1 the set of whose translations is closed L_1. To establish this closure is indeed more than is needed: it is enough to produce for each u among the class of functions to which we may approximate L_1 by polynomials in the translations of our given functions, at least one function whose Fourier transform does not vanish for the argument u.

It may be shown if $F(x)$ is a non-negative function for which

$$\int_{-\infty}^{\infty} F(x)\,dx > A \tag{2.61}$$

and

$$\left[\int_B^{\infty}+\int_{-\infty}^{-B}\right]F(x)\,dx < \varepsilon A, \tag{2.62}$$

then

$$\left|\int_{-\infty}^{\infty} F(x)\,e^{iux}\,dx\right| > \left(\frac{1}{2}-\frac{3\varepsilon}{2}\right)A \qquad \left[|u|\leqq\frac{\pi}{3B}\right]. \tag{2.63}$$

This results from the fact that for $-B\leqq\lambda\leqq B$,

$$\Re\,(e^{iux}) = \cos ux \geqq \cos\frac{\pi}{3} = \frac{1}{2}.$$

Accordingly, if for every value of B and ε, we may find a function of this sort which may be approximated L_1 by polynomials in the translations of $f(x)$, the Fourier transform of $f(x)$ has no zeros. As in the case of Theorem IV, the function $f(x)$ may be replaced by the class Σ, and the class of translations of $f(x)$ by Σ_1. Then the condition that the Fourier transform of $f(x)$ shall have no zeros is to be replaced by the condition that there shall be no zero common to all the Fourier transforms of functions of Σ.

This method is a conceivable one for the investigation of the zeros of functions with known Fourier transforms, and might be applied to the study of the zeros of the Riemann zeta function. So far it has yielded no results. Under certain conditions, the Taylor series may be introduced to reduce the question of the closure of the set of translations of $f(x)$ to that of the closure of the set of functions

$$f^{(n)}(x). \tag{2.65}$$

This matter has not yet, however, been subjected to an adequate investigation.

3. A sub-class of L_1. In connection with Stieltjes distributions of mass, we shall have to consider "kernels" of class L_1, fulfilling a certain more restrictive condition. This condition on a measurable function $f(x)$ is that the series

$$\sum_{k=-\infty}^{\infty} \overline{\lim_{kA+B \leqq x \leqq (k+1)A+B}} |f(x)| \tag{3.01}$$

shall converge. We may easily verify that this condition is in fact independent of A and B, and that it can only be fulfilled by functions of L_1. With it goes a certain definition of the "distance" between two functions $f(x)$ and $g(x)$, this "distance" being defined as

$$(f(x), g(x)) = \sum_{-\infty}^{\infty} \overline{\lim_{k \leqq x \leqq k+1}} |f(x) - g(x)|. \tag{3.02}$$

This notion of "distance" has a very important difference from

$$\int_{-\infty}^{\infty} |f(x) - g(x)|^2 dx \tag{3.03}$$

and

$$\int_{-\infty}^{\infty} |f(x) - g(x)| dx \tag{3.04}$$

in that it is not true that if (3.01) is finite, then

$$\lim_{\epsilon \to 0} \sum_{k=-\infty}^{\infty} \overline{\lim_{k \leqq x \leqq k+1}} |f(x+\epsilon) - f(x)| = 0. \tag{3.05}$$

Accordingly, if we are to prove theorems analogous to those in the last paragraph without a radical change in method of proof, we must introduce a restriction which will make (3.05) hold. Otherwise we shall be unable to establish the analogue of Lemma IIg.

A condition of the desired sort is that $f(x)$ shall be continuous. It will then be uniformly continuous over any finite interval, and we shall be able, first to choose A so large that

$$\sum_{k=0}^{\infty}\left\{\max_{A+k\leq x\leq A+k+1}|f(x)|+\max_{-A-k-1\leq x\leq -A-k}|f(x)|\right\}<\varepsilon \tag{3.06}$$

and then to choose η so small that for $-A-1\leqq x\leqq A+1,\ 0<\eta_1<\eta$,

$$\max|f(x+\eta_1)-f(x)|<\frac{\varepsilon}{2A+2}. \tag{3.07}$$

It will follow that for $0<\eta_1<\eta$,

$$\sum_{k=-\infty}^{\infty}\max_{k\leq x\leq k+1}|f(x+\varepsilon)-f(x)|<3\varepsilon. \tag{3.08}$$

We shall call the class of all continuous functions for which (3.01) is finite, the class M_1, and we shall say that a class Σ of functions of M_1 is *closed* M_1, if whenever $F(x)$ belongs to M_1 and $\varepsilon>0$, there is a polynomial

$$F_1(x)=\sum_{h=1}^{n}A_h f_h(x) \tag{3.09}$$

in functions of the class Σ such that

$$\sum_{k=-\infty}^{\infty}\max_{k\leq x\leq k+1}|F(x)-F_1(x)|<\varepsilon. \tag{3.10}$$

The following theorem is valid:

THEOREM V. *Let $f(x)$ belong to L_1. Then a necessary and sufficient condition that whenever $f_1(x)$ is a function of M_1 and $\varepsilon>0$, there shall exist a function $\alpha(x)$ of M_1 such that*

$$\sum_{k=-\infty}^{\infty}\max_{k\leq x\leq k+1}\left|f_1(x)-\int_{-\infty}^{\infty}f(y)\,\alpha(x-y)\,dy\right|<\varepsilon \tag{3.11}$$

is that for no real u

$$\frac{1}{\sqrt{2\pi}}\int_{-\infty}^{\infty}f(x)\,e^{iux}\,dx=0. \tag{3.12}$$

The proof differs in no essential respect from that of Theorem II. Let us note that if $f_2(x)$ belongs to L_1 and

$$\frac{1}{\sqrt{2\pi}}\int_{-\infty}^{\infty} f_2(x)e^{iux}\,dx = 0, \qquad [|u|>A] \tag{3.13}$$

then $f_2(x)$ belongs to M_1. To see this, let us put

$$\frac{1}{\sqrt{2\pi}}\int_{-\infty}^{\infty} f_2(x)e^{-iux}\,dx = g_2(u); \tag{3.14}$$

$g_2(u)$ is bounded and differs from 0 only over a bounded range, and hence belongs to L_2. Thus

$$\underset{B\to\infty}{\text{l.i.m.}}\ \frac{1}{\sqrt{2\pi}}\int_{-B}^{B} g_2(u)\,e^{iux}\,du = f_3(x) \tag{3.15}$$

exists as a function of L_2. Further

$$g_2(u) = \underset{B\to\infty}{\text{l.i.m.}}\ \frac{1}{\sqrt{2\pi}}\int_{-B}^{B} f_3(x)\,e^{-iux}\,dx \tag{3.16}$$

exists and belongs to L_2. If we put

$$F(x) = \int_{-\infty}^{\infty} G(u)\,e^{iux}\,du \tag{3.17}$$

where G belongs both to L_2 and L_1, it follows readily from (3.14) and (3.16) that

$$\int_{-\infty}^{\infty} F(x)\,[f_2(x)-f_3(x)]\,dx = 0. \tag{3.18}$$

We can choose a set of F's vanishing outside a given finite interval and closed L_2 and hence L_1 over this interval. Then over this interval, which is arbitrary, f_2 and f_3 can differ at most on a null set. Thus f_2 belongs to L_2.

Now

$$g_2(u) = g_2(u)\,W(u) \tag{3.19}$$

where

$$W(u) = \begin{cases} 0\ ; & [|u|>2A] \\ 2-\dfrac{|u|}{A}; & [A<|u|<2A] \\ 1\ . & [|u|<A] \end{cases} \tag{3.20}$$

Thus by the Parseval theorem,

$$\begin{aligned} f_2(x) &= \frac{1}{2\pi}\int_{-\infty}^{\infty} f_2(y)\,dy\int_{-\infty}^{\infty} W(u)\,e^{iu(x-y)}\,du \\ &= \frac{1}{\pi A}\int_{-\infty}^{\infty} f_2(y)\,\frac{\cos A(x-y)-\cos 2A(x-y)}{(x-y)^2}\,dy. \end{aligned} \tag{3.21}$$

From this it follows that

$$(3.22)\quad \sum_{k=-\infty}^{\infty} \max_{k\leq x\leq k+1} |f_2(x)| \leq \frac{1}{\pi A}\int_{-\infty}^{\infty} |f_2(y)|\,dy \sum_{-\infty}^{\infty} O\left(\frac{1}{k-m}\right)^2 \leq O(1)\int_{-\infty}^{\infty} |f_2(y)|\,dy$$

and that $f_2(x)$, which obviously can be modified so as to be continuous, belongs to M_1.

It only remains to prove the analogue of Lemma IIi, and to show that every function of class M_1 may be approximated with any degree of accuracy in the M_1 sense by functions with Fourier transforms vanishing for large arguments. This follows exactly the lines suggested by IIi. We wish to show, that is, that if $f(x)$ belongs to M_1,

$$(3.221)\qquad \lim_{N\to\infty}\left(f(x), \frac{1}{\pi N}\int_{-\infty}^{\infty} f(x+y)\frac{\sin^2 Ny}{y^2}\,dy\right) = 0.$$

As before, we put

$$(3.222)\qquad \Phi_1(y) = \begin{cases} \dfrac{\sin^2 Ny}{y^2}(1-|y|\,N^{1/2}); & [|y| \leq N^{-1/2}] \\ 0 & [y > N^{-1/2}] \end{cases}$$

$$(3.223)\qquad \Phi_2(y) = \frac{\sin^2 Ny}{y^2} - \Phi_1(y).$$

The proof that

$$(3.224)\qquad \lim_{N\to\infty}\left(f(x), \frac{1}{\pi N}\int_{-\infty}^{\infty} f(x+y)\,\Phi_1(y)\,dx\right) = 0$$

follows exactly the lines of that of (2.41), for the analogue of Lemma IIg has already been shown to be true, while that of Lemma IIh is proved by the same arguments of absolute convergence as the lemma itself. Again,

$$(3.225)\quad \lim_{N\to\infty}\sum_{k=-\infty}^{\infty}\max_{k\leq x\leq k+1}\left|\frac{1}{\pi N}\int_{-\infty}^{\infty} f(x+y)\,\Phi_2(y)\,dy\right| \leq 2\sum_{k=-\infty}^{\infty}\max_{k\leq x\leq k+1}|f(x)|\lim_{N\to\infty}\frac{1}{\pi N}\int_{-\infty}^{\infty}\Phi_2(y)\,dy,$$

and this, by (2.421), is zero. This completes the proof of (3.221).

An extension of Theorem V in the direction of Theorem IV is the following:

THEOREM VI. *Let Σ be a class of functions of L_1. Let Σ_1 be the class of all functions of the form*

$$\sum_1^N \int_{-\infty}^{\infty} f_n(y)\, \alpha_n(x+y)\, dy \tag{3.23}$$

where $f_1, \cdots, f_N$ belong to $\sum$ and $\alpha_1, \cdots, \alpha_N$ to M_1. Then Σ_1 is closed M_1 when and only when there is no real value of u which is a zero common to all the functions

$$\frac{1}{\sqrt{2\pi}} \int_{-\infty}^{\infty} f_n(x)\, e^{iux}\, dx \tag{3.24}$$

corresponding to functions $f_n(x)$ of class Σ.

As an easy corollary of Theorem V, using (3.10), we have:

THEOREM VII. *If $f(x)$ is a function of M_1, a necessary and sufficient condition for the set of its translations to be closed M_1 is that for no u should we have*

$$\frac{1}{\sqrt{2\pi}} \int_{-\infty}^{\infty} f(x)\, e^{iux}\, dx = 0. \tag{3.25}$$

CHAPTER II.

ASYMPTOTIC PROPERTIES OF AVERAGES[26].

4. Averages of bounded functions. Our fundamental theorem is:

THEOREM VIII. *Let $f(x)$ be a bounded measurable function, defined over $(-\infty, \infty)$. Let $K_1(x)$ be a function in L_1, and let*

$$\frac{1}{\sqrt{2\pi}} \int_{-\infty}^{\infty} K_1(x)\, e^{iux}\, dx \neq 0 \tag{4.01}$$

for every real u. Let

$$\lim_{x\to\infty} \int_{-\infty}^{\infty} f(\xi)\, K_1(\xi - x)\, d\xi = A \int_{-\infty}^{\infty} K_1(\xi)\, d\xi. \tag{4.02}$$

Then if $K_2(x)$ is any function in L_1,

$$\lim_{x\to\infty} \int_{-\infty}^{\infty} f(\xi)\, K_2(\xi - x)\, d\xi = A \int_{-\infty}^{\infty} K_2(\xi)\, d\xi. \tag{4.03}$$

Conversely, let $K_1(\xi)$ be a function of L_1, and let $\int_{-\infty}^{\infty} K_1(\xi)\, d\xi \neq 0$. Let (4.02) *imply* (4.03) *whenever $K_2(x)$ belongs to L_1 and $f(x)$ is bounded. Then* (4.01) *holds.*

---

[26] The emphasis here placed on averages is in the same order of ideas as was first introduced into the theory by Schmidt (1) (2).

As to the first part of Theorem VIII, it is clear that it is valid whenever $K_2(x)$ is of the form $\sum_{k=1}^{n} A_k K_1(x+\lambda_k)$. It is also clear that if

$$|f(x)| \leqq B; \qquad \int_{-\infty}^{\infty} |K_2(x)-K_3(x)|\, dx \leqq \varepsilon, \tag{4.04}$$

then

$$\left| \int_{-\infty}^{\infty} f(\xi)\, K_2(\xi-x)\, d\xi - \int_{-\infty}^{\infty} f(\xi)\, K_3(\xi-x)\, d\xi \right| \leqq B\varepsilon.$$

An application of Theorem II completes the proof.

As to the second part of the proof, it is merely necessary to suppose that

$$\frac{1}{\sqrt{2\pi}} \int_{-\infty}^{\infty} K_1(x)\, e^{iu_1 x}\, dx = 0 \tag{4.05}$$

and to take

$$f(x) = e^{iu_1 x} \tag{4.06}$$

to obtain a contradiction.

Theorem VIII has an extension in the sense of Theorem III, in which $K_1(x)$ is replaced by a class of functions Σ_1 (4.02) holds for every function of the class, and there is no u for which for *every* function $K_1(x)$ of Σ_1

$$\frac{1}{\sqrt{2\pi}} \int_{-\infty}^{\infty} K_1(x)\, e^{iux}\, dx = 0. \tag{4.07}$$

The converse of this extended theorem is also valid.

5. Averages of bounded Stieltjes distributions. Here our theorem is:

THEOREM IX. *Let $f(x)$ be a function of limited total variation over every finite range, and let*

$$\int_{y}^{y+1} |d f(x)| \tag{5.01}$$

be bounded in y. Let $K_1(x)$ be a continuous function of L_1, and let

$$\sum_{k=-\infty}^{\infty} \overline{\lim_{k \leqq x \leqq k+1}} |K_1(x)| \tag{5.02}$$

converge. Let

$$\frac{1}{\sqrt{2\pi}} \int_{-\infty}^{\infty} K_1(x) e^{iux}\, dx \neq 0 \qquad [-\infty < u < \infty] \tag{4.01}$$

and let

$$\lim_{x\to\infty} \int_{-\infty}^{\infty} K_1(\xi - x)\, df(\xi) = A \int_{-\infty}^{\infty} K_1(\xi)\, d\xi. \tag{5.03}$$

Then if $K_2(x)$ is any function of M_1,

$$\lim_{x\to\infty} \int_{-\infty}^{\infty} K_2(\xi - x)\, df(\xi) = A \int_{-\infty}^{\infty} K_2(\xi)\, d\xi. \tag{5.04}$$

Conversely, let (5.02) *converge. Let* $\int_{-\infty}^{\infty} K_1(x)\,dx \neq 0$. *Let* (5.03) *imply* (5.04) *for every function of* K_2 *of* M_1 *and every* $f(x)$ *for which* (5.01) *is bounded. Then* (4.01) *holds.*

To prove this theorem we need the following elementary:

LEMMA IXa. *Let*

$$\int_{y-0}^{y+1+0} |df(x)| \leq B \qquad [-\infty < y < \infty] \tag{5.05}$$

and let

$$\sum_{k=-\infty}^{\infty} \overline{\lim_{k \leq x \leq k+1}} |K_2(x) - K_3(x)| \leq \epsilon. \tag{5.06}$$

Then

$$\left| \int_{-\infty}^{\infty} K_2(\xi - x)\,df(x) - \int_{-\infty}^{\infty} K_3(\xi - x)\,df(x) \right| \leq B\epsilon. \tag{5.061}$$

With this lemma at our disposal, the proof of Theorem IX does not differ in any important respect from that of Theorem VIII. Of course, as in Theorem VI, we must replace the polynomials in translations of K_1 that figure in Theorem VIII by absolutely convergent integrals in these translations. The proof of the lemma itself is immediate.

The modification of Theorem IX with a hypothesis involving a whole class of kernels will be used later, so we shall formulate it as a separate theorem:

THEOREM X. *Let* $f(x)$ *be a function of limited total variation over every finite range, and let*

$$\int_{y}^{y+1} |df(x)| \tag{5.01}$$

be bounded in y. *Let* Σ *be a class of continuous functions of* L_1, *each one of which, for example* $K_1(x)$, *has the properties that*

$$\sum_{k=-\infty}^{\infty} \overline{\lim_{k \leq x \leq k+1}} |K_1(x)| \tag{5.02}$$

converges, and that

$$\lim_{x\to\infty} \int_{-\infty}^{\infty} K_1(\xi - x)\,df(\xi) = A \int_{-\infty}^{\infty} K_1(\xi)\,d\xi. \tag{5.03}$$

Let there be no u *which is a real zero for all the functions*

$$\frac{1}{\sqrt{2\pi}} \int_{-\infty}^{\infty} K_1(x)\,e^{iux}\,dx \tag{5.07}$$

for which $K_1(x)$ *belongs to* Σ. *Then if* $K_2(x)$ *is any function belonging to* M_1,

$$(5.04)\qquad \lim_{x\to\infty}\int_{-\infty}^{\infty} K_2(\xi-x)\,df(\xi) = A\int_{-\infty}^{\infty} K_2(\xi)\,d\xi.$$

Conversely, if the class $\sum$ *contains at least one function with non zero integral and has the property that each member of it is a member of* L_1 *and satisfies the condition that* (5.02) *shall converge, and if, whenever* $f(x)$ *satisfies the condition that* (5.01) *be bounded, and* (5.03) *for every member of* Σ, *then* (5.04) *holds for every function* $K_2(x)$ *belonging to* M_1, *then there is no* u *such that* (4.07) *holds for every function* $K_1(x)$ *belonging to* Σ.

6. **Averages of unilaterally bounded distributions and functions.** Let $K_1(x)$ not be equivalent to 0, and let it be a continuous function (or a function continuous except for a finite number of finite jumps) not identically zero and such that

$$(6.01)\qquad K_1(x)\geqq 0,\quad \sum_{k=-\infty}^{\infty}\max_{k\leqq x\leqq k+1} K_1(x)<p\qquad(-\infty<x<\infty).$$

We shall say that the mass-distribution determined by $f(x)$ is *bounded below* when

$$(6.02)\qquad \int_y^{y+1}|df(x)|-f(y+1)+f(y)\leqq N\qquad(-\infty<x<\infty)$$

and that it is *bounded above* when $-f(x)$ determines a mass-distribution bounded below. We shall prove:

LEMMA XIa. *Let the distribution corresponding to* $f(x)$ *be bounded below (or above). Let* (6.01) *hold, and let*

$$(6.03)\qquad \overline{\int_{-\infty}^{\infty}} K_1(\xi-x)\,df(\xi)\leqq M;\quad(\geqq M);\qquad(-\infty<x<\infty).$$

Then there is a Q *such that*

$$(6.04)\qquad \int_y^{y+1}|df(x)|\leqq Q\qquad(-\infty<y<\infty).$$

To prove this, let us notice that

$$(6.05)\qquad \begin{aligned}\overline{\int_{-\infty}^{\infty}} K_1(\xi-x)\,d\int_{\xi_0}^{\xi}|df(y)| &\leqq \overline{\int_{-\infty}^{\infty}} K_1(\xi-x)\,d\left[\int_{\xi_0}^{\xi}|df(y)|-f(\xi)\right]\\ &+\overline{\int_{-\infty}^{\infty}} K_1(\xi-x)\,df(\xi)\leqq N\sum_{k=-\infty}^{\infty}\max_{k\leqq x\leqq k+1}K(\xi)+M.\end{aligned}$$

Since K is "stückweise stetig", B, a and b exist such that

$$(6.06)\qquad K_1(u)>B>0\qquad(a\leqq u\leqq b).$$

Hence

$$(6.07)\qquad M+N\sum_{k=-\infty}^{\infty}\max_{k\leqq x\leqq k+1}K_1(\xi)\geqq B\int_{a+x}^{b+x}d\int_{\xi_0}^{\xi}|df(y)|,$$

or in other terms

$$(6.08)\qquad \frac{1}{B}\left\{\left[\frac{1}{b-a}\right]+1\right\}\left\{M+N\sum_{k=-\infty}^{\infty}\max_{k\leqq x\leqq k+1}K_1(\xi)\right\} \geqq \int_y^{y+1}|df(x)| \qquad (-\infty<y<\infty).$$

If we combine this lemma with Theorem X, we get:

THEOREM XI. *Let $f(x)$ be a function of limited total variation over every finite interval, for which*

$$(6.02)\qquad \int_y^{y+1}|df(x)|-f(y+1)+f(y)\leqq N \qquad (-\infty<y<\infty).$$

Let Σ be a class of continuous functions K_1, for each of which

$$(5.02)\qquad \sum_{k=-\infty}^{k=\infty}\max_{k\leqq x\leqq k+1}|K_1(x)|$$

converges, and let

$$(5.03)\qquad \lim_{x\to\infty}\int_{-\infty}^{\infty}K_1(\xi-x)\,df(\xi)=A\int_{-\infty}^{\infty}K_1(\xi)\,d\xi$$

for every K_1 belonging to Σ. Let there be no real u for which every

$$(5.07)\qquad \frac{1}{\sqrt{2\pi}}\int_{-\infty}^{\infty}K_1(x)\,e^{iux}\,dx$$

vanishes, for which K_1 belongs to Σ. Let $Q(x)$ be a continuous function belonging to Σ, for which

$$(6.09)\qquad Q(x)\geqq 0, \qquad \left|\int_{-\infty}^{\infty}Q(\xi-x)\,df(x)\right|\leqq T \qquad (-\infty<x<\infty).$$

Then if $K_2(-x)$ is any function belonging to M_1,

$$(5.04)\qquad \lim_{x\to\infty}\int_{-\infty}^{\infty}K_2(\xi-x)\,df(\xi)=A\int_{-\infty}^{\infty}K_2(\xi)\,d\xi.$$

We shall have frequent occasion to use Theorems X and XI in a form in which the infinite range of x is mapped on a semi-infinite range by an exponential transformation. This may happen in two ways: either 0 or ∞ for the new argument may correspond to $+\infty$ for x. In the first case, let us write:

$$(6.10)\qquad \xi=-\log\lambda,\quad f(\xi)=\int\lambda^{-1}\,d\varphi(\lambda),\quad K_{1,2}(\xi)=\lambda N_{1,2}(\lambda),\quad Q(\xi)=\lambda M(\lambda)$$

and in the second,

$$(6.11)\qquad \xi=\log\lambda,\quad f(\xi)=\int\lambda^{-1}\,d\varphi(\lambda),\quad K_{1,2}(\xi)=\lambda N_{1,2}(\lambda),\quad Q(\xi)=\lambda M(\lambda).$$

Then Theorem XI yields:

THEOREM XI′. *Let $\varphi(\lambda)$ be a function of limited total variation over every interval $(\varepsilon, 1/\varepsilon)$ where $0<\varepsilon<1$. Let $\varphi(0)=0$ and let*

$$(6.12) \qquad \int_u^{2u} \lambda^{-1} \, |d\varphi(\lambda)| - \int_u^{2u} \lambda^{-1} \, d\varphi(\lambda) \leqq N \qquad (0<u<\infty).$$

Let Σ be a class of continuous functions $N_1(\lambda)$ for each of which

$$(6.13) \qquad \sum_{k=-\infty}^{\infty} \max_{2^k \leqq \lambda \leqq 2^{k+1}} \lambda \, |N_1(\lambda)|$$

converges, and let

$$(6.14) \qquad \lim_{\lambda \to 0(\infty)} \frac{1}{\lambda} \int_0^\infty N_1\left(\frac{\mu}{\lambda}\right) d\varphi(\mu) = A \int_0^\infty N_1(\mu)\, d\mu$$

for every $N_1(\lambda)$ belonging to Σ. Let there be no real u for which every

$$(6.15) \qquad \int_0^\infty N_1(\lambda)\, \lambda^{iu}\, d\lambda$$

vanishes, for which N_1 belongs to Σ. Let $M(\lambda)$ be a function belonging to Σ, for which

$$(6.16) \qquad M(\lambda) \geqq 0, \quad \left| \frac{1}{\lambda} \int_0^\infty M\left(\frac{\mu}{\lambda}\right) d\varphi(\mu) \right| \leqq \text{const.}\ \textit{for}\ 0 \leqq \lambda < \infty.$$

Then if $N_2(\lambda)$ is any continuous function for which

$$(6.17) \qquad \sum_{k=-\infty}^{\infty} \max_{2^k \leqq \lambda \leqq 2^{k+1}} \lambda \, |N_2(\lambda)|$$

converges,

$$(6.18) \qquad \lim_{\lambda \to 0(\infty)} \frac{1}{\lambda} \int_0^\infty N_2\left(\frac{\mu}{\lambda}\right) d\varphi(\mu) = A \int_0^\infty N_2(\mu)\, d\mu.$$

In the case where

$$(6.19) \qquad \varphi(\lambda) = \int_{\lambda_0}^{\lambda} f(x)\, dx,$$

(6.12) will be satisfied if

$$(6.20) \qquad f(\lambda) > -\frac{N}{2 \log 2}.$$

If the other hypotheses of Theorem XI′ are satisfied, (6.18) will become

$$(6.21) \qquad \lim_{\lambda \to 0(\infty)} \frac{1}{\lambda} \int_0^\infty N_2\left(\frac{\mu}{\lambda}\right) f(\mu)\, d\mu = A \int_0^\infty N_2(\mu)\, d\mu.$$

As a particular admissible N_2, we may take

$$(6.22)\qquad N_2(\mu) = \begin{cases} 1 & ; \quad [0 \leqq \mu < 1] \\ 1 - \dfrac{\mu-1}{\varepsilon}; & [1 \leqq \mu < 1+\varepsilon] \\ 0 & . \quad [1+\varepsilon \leqq \mu] \end{cases}$$

Thus (6.21) will assume the form

$$(6.23)\qquad \lim_{\lambda\to 0(\infty)} \frac{1}{\lambda}\left[\int_0^\lambda f(\mu)\,d\mu + \int_\lambda^{\lambda(1+\varepsilon)} \left(1-\frac{\mu-\lambda}{\lambda\varepsilon}\right) f(\mu)\,d\mu\right] = A\left(1+\frac{\varepsilon}{2}\right).$$

Now, by (6.20)

$$(6.24)\qquad \frac{1}{\lambda}\int_\lambda^{\lambda(1+\varepsilon)} \left(1-\frac{\mu-\lambda}{\lambda\varepsilon}\right) f(\mu)\,d\mu \geqq \frac{1}{\lambda}\int_\lambda^{\lambda(1+\varepsilon)} \left(1-\frac{\mu-\lambda}{\lambda\varepsilon}\right)\left(\frac{-N}{2\log 2}\right)d\mu$$
$$= \frac{-N\varepsilon}{4\log 2}.$$

Hence by (6.23)

$$(6.25)\qquad \overline{\lim_{\lambda\to 0(\infty)}} \frac{1}{\lambda}\int_0^\lambda f(\mu)\,d\mu \leqq A\left(1+\frac{\varepsilon}{2}\right) + \frac{N\varepsilon}{4\log 2}.$$

Since ε is arbitrarily small,

$$(6.25)\qquad \overline{\lim_{\lambda\to 0(\infty)}} \frac{1}{\lambda}\int_0^\lambda f(\mu)\,d\mu \leqq A.$$

Again, we may write (6.21) in the form

$$(6.26)\qquad \lim_{\lambda\to 0(\infty)} \frac{1}{\lambda(1+\varepsilon)}\left[\int_0^{\lambda(1+\varepsilon)} f(\mu)\,d\mu - \int_\lambda^{\lambda(1+\varepsilon)} \frac{\mu-\lambda}{\lambda\varepsilon} f(\mu)\,d\mu\right] = \frac{A(1+\varepsilon/2)}{1+\varepsilon}$$

from which we may conclude as above that

$$(6.27)\qquad \underline{\lim_{\lambda\to 0(\infty)}} \frac{1}{\lambda}\int_0^\lambda f(\mu)\,d\mu \geqq A.$$

Combining (6.25) and (6.27), we see that

$$(6.28)\qquad \lim_{\lambda\to 0(\infty)} \frac{1}{\lambda}\int_0^\lambda f(\mu)\,d\mu = A.$$

We thus get

THEOREM XI″. *Let $f(x)$ be a function bounded over every interval $(\varepsilon, 1/\varepsilon)$, where $0 < \varepsilon < 1$. Let*

$$(6.29)\qquad f(\lambda) > -K \ (or\ f(\lambda) < K)$$

for every argument. Let Σ be a class of continuous functions $N_1(\lambda)$ for each of which (6.13) *converges, and let*

$$(6.30)\qquad \lim_{\lambda\to 0(\infty)} \frac{1}{\lambda}\int_0^\infty N_1\left(\frac{\mu}{\lambda}\right) f(\mu)\,d\mu = A\int_0^\infty N_1(\mu)\,d\mu$$

for every $N_1(\lambda)$ belonging to Σ. Let there be no real u for which every expression (6.15) *vanishes with N_1 belonging to Σ. Let $M(\lambda)$ belong to Σ, and let*

$$(6.31)\qquad M(\lambda) \geqq 0,\ \frac{1}{\lambda}\int_0^\infty M\left(\frac{\mu}{\lambda}\right) f(\mu)\,d\mu \leq \text{const. } for\ 0 \leqq \lambda < \infty .$$

Then (6.28) *is true.*

Another case where a stricter conclusion than (6.18) may be drawn is where $\varphi(\lambda)$ is monotone. In this case, (6.12) is automatically satisfied. If we take $N_2(\mu)$ as in (6.22), we shall be able to write (6.18) in the form

$$(6.32)\qquad \lim_{\lambda\to 0(\infty)} \frac{1}{\lambda}\left[\varphi(\lambda) + \int_\lambda^{\lambda(1+\varepsilon)} \left(1 - \frac{\mu-\lambda}{\lambda\varepsilon}\right) d\varphi(\mu)\right] = A\left(1+\frac{\varepsilon}{2}\right)$$

and hence

$$(6.33)\qquad \overline{\lim_{\lambda\to 0(\infty)}} \frac{\varphi(\lambda)}{\lambda} \leqq A.$$

From the analogue of (6.26) it follows that

$$(6.34)\qquad \underline{\lim_{\lambda\to 0(\infty)}} \frac{\varphi(\lambda)}{\lambda} \geqq A,$$

and hence that

$$(6.35)\qquad \varphi(\lambda) \sim \lambda A.$$

We thus obtain

THEOREM XI‴. *If in the hypothesis of Theorem* XI′, (6.12) *is replaced by the condition that $\varphi(\lambda)$ is monotone. Then* (6.35) *follows.*

It is even possible to weaken the hypothesis here given, and to replace (6.12) by the condition that

$$(6.36)\qquad \lim_{\varepsilon\to +0}\ \overline{\lim_{-\infty<\mu<\infty}} \left[\int_\mu^{\mu(1+\varepsilon)} \lambda^{-1}\,|d\varphi(\lambda)| - \int_\mu^{\mu(1+\varepsilon)} \lambda^{-1}\,d\varphi(\lambda)\right] = 0$$

to establish (6.25).

CHAPTER III.

TAUBERIAN THEOREMS AND THE CONVERGENCE OF SERIES AND INTEGRALS.

7. **The Hardy-Littlewood condition.** We now enter upon the realm of ideas which has longest been associated with Tauberian theorems.[27] The class Σ of the last paragraph now consists of all functions of the form

[27] Hardy (2); Hardy and Littlewood (4), (10), (13) etc.; Littlewood (1); Landau (9), (4).

(7.01) $$M(x) - M(x+\lambda)$$

for which we have

(7.02) $$M(-\infty) = 1, \quad M(\infty) = 0, \quad \int_{-\infty}^{\infty} M(\xi - x)\, df(\xi) \quad \text{bounded.}$$

Condition (5.02) now asserts the convergence of

(7.03) $$\sum_{k=-\infty}^{\infty} \max_{k \leqq x \leqq k+1} |M(x) - M(x+\lambda)|$$

which will be automatically fulfilled if $M(x)$ is monotone and satisfies (7.02). We shall also suppose $M(x)$ continuous. Condition (5.07) becomes the condition that

(7.04) $$\frac{1}{\sqrt{2\pi}} \int_{-\infty}^{\infty} [M(x) - M(x+\lambda)]\, e^{iux}\, dx$$

shall not vanish for any real u for every λ. Let us suppose M to vanish exponentially at $+\infty$, and let

(7.05) $$\frac{1}{\sqrt{2\pi}} \int_{-\infty}^{\infty} M(x)\, e^{xz}\, dx = \varphi(z).$$

$\varphi(z)$ will clearly be analytic over some vertical strip to the right of the origin. We shall obviously have

(7.06) $$\frac{1}{\sqrt{2\pi}} \int_{-\infty}^{\infty} [M(x) - M(x+\lambda)]\, e^{xz}\, dz = \varphi(z)\,(1 - e^{-\lambda z})$$

over this strip, and by analytic continuation, which we assume to be possible, the non-vanishing of (7.04) becomes the non-vanishing of $\varphi(z)$ over the imaginary axis.

If

(7.07) $$\lim_{x \to \infty} \int_{-\infty}^{\infty} M(\xi - x)\, df(\xi) = B,$$

(5.03) is satisfied for $A = 0$. As is obviously permissible, let us put

(7.08) $$K_2(\xi) = \begin{cases} M(\xi) - 1 \quad ; & (-\infty < \xi < 0) \\ M(\xi) - 1 + \dfrac{\xi}{\varepsilon}; & (0 \leqq \xi < \varepsilon) \\ M(\xi) \quad . & (\varepsilon \leqq \xi < \infty) \end{cases}$$

Then (5.04) assumes the form

(7.09) $$\lim_{x \to \infty} \int_{-\infty}^{\infty} K_2(\xi - x)\, df(\xi) = 0,$$

which becomes

(7.10) $$\lim_{x \to \infty} \left[\int_{-\infty}^{x} df(\xi) + \int_{x}^{x+\varepsilon} \left(1 - \frac{\xi - x}{\varepsilon}\right) df(\xi) \right] = B.$$

If we put

(7.11) $$f(-\infty) = 0$$

(5.04) assumes the form

(7.12) $$\lim_{x\to\infty} \frac{1}{\varepsilon} \int_x^{x+\varepsilon} f(\xi)\, d\xi = B.$$

Restating Theorem XI in this new form, we get

THEOREM XII. *Let $f(x)$ be a function of limited total variation over every range $(-\infty, A)$, let $f(-\infty) = 0$, and let*

(6.02) $$\int_y^{y+1} |df(x)| - f(y+1) + f(y) \leqq N \qquad (-\infty < y < \infty).$$

Let $M(x)$ be a monotonely decreasing continuous function, such that

(7.13) $$M(-\infty) = 1, \quad M(x) = O(e^{-ux}) \text{ at } \infty \qquad (u > 0).$$

Let

(7.05) $$\frac{1}{\sqrt{2\pi}} \int_{-\infty}^{\infty} M(x)\, e^{xz}\, dx = \varphi(z)$$

over a strip to the right of the origin, and let $\varphi(z)$, continued analytically on to the imaginary axis, have no zeros there. (We assume the possibility of this continuation.) *Let*

(7.14) $$\int_{-\infty}^{\infty} M(\xi - x)\, df(\xi)$$

be bounded, and let

(7.07) $$\lim_{x\to\infty} \int_{-\infty}^{\infty} M(\xi - x)\, df(\xi) = B.$$

Then

(7.12) $$\lim_{x\to\infty} \frac{1}{\varepsilon} \int_x^{x+\varepsilon} f(\xi)\, d\xi = B.$$

Let us put

$$M(\log x) = N(x); \quad f(\log x) = F(x).$$

Theorem XII then becomes:

THEOREM XIII. *Let $F(x)$ be a function vanishing at the origin and of limited total variation over any finite range (including the origin), and let for some $A > 1$,*

(7.16) $$\int_y^{Ay} |dF(x)| - F(Ay) + F(y) \leqq N \qquad (0 < y < \infty).$$

Let $N(x)$ be a monotonely decreasing continuous function, such that

(7.17) $$N(0) = 1, \qquad N(x) = O(x^{-\lambda}) \text{ at } \infty \qquad [\lambda > 0].$$

Let

(7.18) $$\frac{1}{\sqrt{2\pi}} \int_0^{\infty} N(x)\, x^{z-1}\, dx = \varphi(z)$$

over a strip to the right of the origin, and let $\varphi(z)$, continued analytically on to the imaginary axis, as we assume to be possible for all points but the origin, have no zeros there. Let

$$\int_0^\infty N\left(\frac{\xi}{x}\right) dF(\xi) \tag{7.19}$$

be bounded, and let

$$\lim_{x\to\infty} \int_0^\infty N\left(\frac{\xi}{x}\right) dF(\xi) = B. \tag{7.20}$$

Then if $\Lambda_1 > 1$,

$$\lim_{x\to\infty} \frac{1}{\log \Lambda_1} \int_x^{x\Lambda_1} F(\xi) \frac{d\xi}{\xi} = B. \tag{7.21}$$

A case of particular interest and importance is where

$$F(x) = \sum_{n=0}^{[x]} a_n. \tag{7.22}$$

Here

$$\begin{aligned} \int_y^{\Lambda y} |dF(x)| - F(\Lambda y) + F(y) &= \sum_{n=[y]+1}^{[\Lambda y]} (-a_n + |a_n|) \\ &\leqq \begin{cases} \max 2 n a_n; \ (n \leqq [\Lambda y]), \\ 0. \end{cases} \end{aligned} \tag{7.23}$$

if Λ is sufficiently near to 1. (7.23) is to be interpreted as meaning that the expression on the left hand side is less than or equal to the greater of the two expressions in brackets.

We thus obtain the

COROLLARY. *If $N(x)$ is subject to the conditions of Theorem* XIII, *if*

$$\int_0^\infty N\left(\frac{\xi}{x}\right) dF(\xi) = \sum_{n=0}^\infty a_n N\left(\frac{n}{x}\right) \tag{7.24}$$

is bounded, if

$$n a_n < K \qquad (\text{or } n a_n > -K) \tag{7.25}$$

for all n, and if

$$\lim_{x\to\infty} \sum_{n=0}^\infty a_n N\left(\frac{n}{x}\right) = B \tag{7.26}$$

then

$$\sum_0^\infty a_n = B. \tag{7.27}$$

Here instead of (7.27), what we directly prove is

$$\lim_{x\to\infty} \frac{1}{\log \Lambda_1} \int_x^{x\Lambda_1} \sum_{n=0}^{[\xi]} a_n \frac{d\xi}{\xi} = B. \tag{7.28}$$

(7.27) may however be written in the form

$$\lim_{x\to\infty} \frac{1}{\log A_1}\int_x^{xA_1} \sum_{n=0}^{[x]} a_n \frac{d\xi}{\xi} = B. \tag{6.281}$$

Our conclusion (7.27) will then follow if we can show that we can so choose A_1, that

$$\begin{aligned} &\overline{\lim_{x\to\infty}} \frac{1}{\log A_1}\int_x^{xA_1} \sum_{n=[x]+1}^{[\xi]} a_n \frac{d\xi}{\xi} < \varepsilon, \\ &\overline{\lim_{x\to\infty}} \frac{1}{\log A_1}\int_x^{xA_1} \sum_{n=[\xi]+1}^{[xA_1]} a_n \frac{d\xi}{\xi} < \varepsilon. \end{aligned} \tag{7.29}$$

However,

$$\frac{1}{\log A_1}\int_x^{xA_1} \sum_{n=[x]+1}^{[\xi]} a_n \frac{d\xi}{\xi} \leq \frac{1}{\log A_1}\int_x^{xA_1} \frac{K}{x}(xA_1 - x)\frac{d\xi}{\xi} = K(A_1 - 1). \tag{7.30}$$

Again,

$$\frac{1}{\log A_1}\int_x^{xA_1} \sum_{n=[\xi]+1}^{[A_1 x]} a_n \frac{d\xi}{\xi} \leq K(A_1 - 1). \tag{7.31}$$

Since we may take A_1 as near to 1 as we wish, (7.29) is established, and (7.27) follows at once. This type of argument is to be found in the work of Szász.

8. **The Schmidt condition.**[28] This last corollary covers the work of Hardy and Littlewood on Tauberian theorems. An extension of their conditions is due to Robert Schmidt. To arrive at theorems of his type, let M be subject to the conditions stated in the hypothesis to Theorem XII; let us put

$$g(x) = f(x+\alpha) - f(x) \tag{8.01}$$

and let us assume that

$$-f(x) + f(x) < A(x + |x|) + c \qquad (A, c > 0). \tag{8.02}$$

As the integrals in question converge absolutely by (8.02) and (7.13), whenever they exist we may invert our order of integration and write:

$$\begin{aligned} \int_y^{y+\alpha} dx \int_{-\infty}^{\infty} M(\xi - x)\, df(\xi) &= \int_y^{y+\alpha} dx \int_{-\infty}^{\infty} M(\xi)\, df(\xi + x) \\ &= \int_{-\infty}^{\infty} M(\xi)\big(f(\xi+y+\alpha) - f(\xi+y)\big)\, d\xi \\ &= \int_{-\infty}^{\infty} M(\xi - y)\big(f(\xi+\alpha) - f(\xi)\big)\, d\xi \\ &= \int_{-\infty}^{\infty} M(\xi - y)\, g(\xi)\, d\xi. \end{aligned} \tag{8.03}$$

[28] Schmidt (1), (2). Throughout this section, the author has been strongly influenced by the methods of Vijayaraghavan and Szász.

Thus if $M(x)$ is subject to the conditions in the hypothesis to Theorem XII, if it is bounded above or below, and (7.07) is valid, and (7.14) bounded, whether (6.02) is true or not, it follows from Theorem XII itself on replacing $f(x)$ by $\frac{1}{\alpha}\int_{-\infty}^{x} g(\xi)\,d\xi$ that

$$
\begin{aligned}
B &= \lim_{y\to\infty} \frac{1}{\alpha}\int_{y}^{y+\alpha} \frac{dx}{\alpha}\int_{-\infty}^{x} g(\xi)\,d\xi \\
&= \lim_{y\to\infty} \frac{1}{\alpha}\int_{y}^{y+\alpha} \frac{dx}{\alpha}\int_{x}^{x+\alpha} f(\xi)\,d\xi \qquad (8.04)\\
&= \lim_{y\to\infty} \frac{1}{\alpha^2}\int_{-\alpha}^{\alpha} f(\xi+\alpha+y)\,(\alpha-|\xi|)\,d\xi.
\end{aligned}
$$

Let us now introduce Schmidt's notion of a "slowly decreasing" function. Let $f(u)$ have the property that when u and v run through a sequence of pairs of values for which

$$v \geqq u, \qquad v-u \to 0 \tag{8.05}$$

then

$$\varliminf_{u\to\infty} (f(v)-f(u)) \geqq 0; \tag{8.06}$$

we shall call $f(u)$ *slowly decreasing*. (Schmidt treats sequences instead of functions, and on another scale, but the difference is unimportant.) Schmidt shows that we may write

$$f(u) = f_1(u)+f_2(u) \tag{8.07}$$

where $f_2(u)$ is monotone increasing, and $f_1(u)$ satisfies the condition that whenever (8.05) is fulfilled, then

$$\lim_{u\to\infty} (f_1(v)-f_1(u)) = 0. \tag{8.08}$$

He proves that there exists a function $T(\alpha)$ such that

$$\varlimsup_{\substack{u\to\infty\\ u\leqq v\leqq u+\alpha}} |f_1(v)-f_1(u)| = T(\alpha) \tag{8.09}$$

and

$$\lim_{\alpha\to 0} T(\alpha) = 0. \tag{8.10}$$

He proves that a number T exists, such that

$$|f_1(v)-f_1(u)| \leqq T\cdot(v-u) \qquad (v>u+\alpha). \tag{8.11}$$

From this, it readily follows that if f is bounded near $-\infty$, (8.02) is valid and $g(x)$ is bounded below. Furthermore,

$$\varlimsup_{x\to\infty} \left(f(x)-\frac{1}{\alpha^2}\int_{-\alpha}^{\alpha} f(\xi+\alpha+x)\,(\alpha-|\xi|)\,d\xi\right) \leqq T(2\alpha) \tag{8.12}$$

and

$$(8.13)\quad \underline{\lim}_{x\to\infty}\left(-\frac{1}{\alpha^2}\int_{-\alpha}^{\alpha} f(\xi+\alpha+x)\,(\alpha-|\xi|)\,d\xi+f(x+2\alpha)\right)\geqq -T(2\alpha);$$

going to the limit

$$(8.14)\quad \underline{\lim}_{x\to\infty} f(x)\geqq -T(2\alpha)+B$$

and

$$(8.15)\quad \overline{\lim}_{x\to\infty} f(x)\leqq B+T(2\alpha).$$

Then by (8.10), it follows that

$$(8.16)\quad \underline{\lim}_{x\to\infty} f(x) = B.$$

This yields

THEOREM XIV. *In the hypothesis of Theorem* XII, *condition* (6.02) *may be replaced by the assumption that* $f(x)$ *is of limited total variation over any finite range, and is slowly decreasing, and the conclusion* (7.12) *may be replaced by* (8.16). Again, Theorem XIII becomes

THEOREM XV. *In the hypothesis to Theorem* XIII, *condition* (7.16) *may be replaced by the condition that if* u *and* v *run through a sequence for which*

$$(8.17)\quad v\geqq u,\quad \frac{v}{u}\to 1$$

then

$$(8.18)\quad \underline{\lim}_{u\to\infty}\,(F(v)-F(u))\geqq 0.$$

In the conclusion to Theorem XIII, (7.21) *may then be replaced by*

$$(8.161)\quad \lim_{x\to\infty} F(x) = B.$$

In the corollary to Theorem XIII, (7.25) *may be replaced by the hypothesis that if*

$$(8.19)\quad s_n = a_0+a_1+\cdots+a_n$$

and $p\to\infty$, $q\to\infty$ *in such a manner that*

$$(8.20)\quad q>p,\quad \frac{q}{p}\to 1$$

then

$$(8.21)\quad \underline{\lim}_{p\to\infty}\,(s_q-s_p)\geqq 0.$$

The conclusion of this corollary remains unchanged.

The hypothesis that (8.20) shall imply (8.21) is Schmidt's hypothesis concerning slowly decreasing sequences in unaltered form.

It will be noted that the proof here given for the Schmidt theorems does not essentially differ from the proofs given by Vijayaraghavan[29] and Szász[30] for theorems of these types. In both cases the transition is made to a Tauberian theorem of more standard type in which s_n is itself averaged, instead of appearing in the conclusion as the average of a unilaterally or bilaterally bounded mass distribution. In both cases, moreover, the Schmidt condition is again used to give a bilateral or unilateral estimate of the difference between s_n and its average.

CHAPTER IV.

TAUBERIAN THEOREMS AND PRIME NUMBER THEORY.

9. Tauberian theorems and Lambert series. One of the most important applications of Tauberian theorems is to the proof of the prime number theorem of Hadamard[31] and de la Vallée Poussin,[32] to the effect that the number of primes less than N is asymptotically $N/\log N$. The prime numbers bear an exceedingly close relation to series of the form

$$\sum a_n \frac{x^n}{1-x^n} \tag{9.01}$$

known as Lambert series, after their eighteenth century discoverer. Until recent times, however, all attempts to employ Lambert series effectively in the study of prime numbers had proved a failure, and indeed Knopp[33] has characterized one of these directions of attack as "verführerisch". Hardy and Littlewood[34] finally showed that the prime number theorem was equivalent to a Tauberian theorem concerning Lambert series, but did not succeed in establishing an autonomous proof of this theorem.

Our general Tauberian theorems suffice to furnish this autonomous proof, and indeed, the Tauberian theorem which we shall find it easiest to establish directly leads more directly to the prime number theorem than does the theorem of Hardy and Littlewood. The latter is also directly demonstrable by our methods. In both, the cardinal point in the proof is that the Riemann zeta function, $\zeta(x+iy)$, has no zeros on the line $x=1$. The proof of this goes back to the first proofs of the prime number theorem, and has always been recognized as that property of the Riemann zeta

[29] Vijayaraghavan (1), (2).

[30] Szász (1), (2).

[31] Hadamard (1).

[32] de la Vallée Poussin (1).

[33] Knopp (3).

[34] Hardy and Littlewood (11); Hardy (6); Ananda-Rau (2).

function which is most central in the proofs of this theorem, but all earlier proofs had made some use of the behavior of the zeta function at infinity. These further properties now appear as inessential, and the non-vanishing of the function becomes the only non-elementary feature of the zeta function in question.

For example, the usual proof of the prime number theorem employs a lemma of Landau[35] to the following effect:

LANDAU'S LEMMA. *Let*

$$F(x) = \sum_{n=1}^{\infty} a_n n^{-x} \qquad [\Re(x) > 1] \tag{9.02}$$

and let

$$a_n \geqq 0 \qquad [n = 1, 2, \cdots]. \tag{9.03}$$

Let $F(x)$ when analytically continued be without singularities on $\Re(x) = 1$, except for a pole of order one at $x = 1$, with principal part $A/(x-1)$. Let there be some α for which

$$F(x) = O(|x|^\alpha) \tag{9.04}$$

in the right half-plane. Then

$$A = \lim_{n \to \infty} \frac{1}{n} \sum_{1}^{n} a_k. \tag{9.05}$$

In the following section, we shall show—following Ikehara— that condition (9.04) is not needed. When the lemma is used in the proof of the prime number theorem,

$$a_n = \Lambda(n) \text{ (to be defined immediately)}; \quad F(x) = -\zeta'(x)/\zeta(x) \tag{9.06}$$

and it will be seen that here too the only non-elementary property of the zeta function which we use is that it does not vanish on the line $\Re(x) = 1$.

To return to Lambert series, let $\Lambda(n)$ be the number-theoretic function determined by:

$\Lambda(p^k) = \log p$ if p is a prime and k is a positive integer;

$\Lambda(n) = 0$ if n is not of the form p^k.

Let $0 \leqq x < 1$; then

$$\sum_{1}^{\infty} \log m\, x^m = \sum_{m=1}^{\infty} x^m \sum_{n/m} \Lambda(n) = \sum_{n=1}^{\infty} \Lambda(n) \sum_{n/m} x^m = \sum_{n=1}^{\infty} \Lambda(n) \frac{x^n}{1-x^n}. \tag{9.07}$$

[35] Landau (1).

Hence

$$
\begin{aligned}
\sum_1^\infty \Lambda(n)\,\frac{x^n}{1-x^n} &= \sum_1^\infty \log m\,\frac{x^m-x^{m+1}}{1-x}\\
&= \frac{x}{1-x}\sum_1^\infty \log\left(1+\frac{1}{m}\right)x^m\\
&= \frac{x}{1-x}\sum_1^\infty\left[\frac{1}{m}+O\left(\frac{1}{m^2}\right)\right]x^m\\
&= \frac{x}{1-x}\left[\log\frac{1}{1-x}+\sum_1^\infty O\left(\frac{1}{m^2}\right)x^m\right].
\end{aligned}\tag{9.08}
$$

If we put $x = e^{-\xi}$ and multiply by ξ we have

$$
\sum_1^\infty \Lambda(n)\,\frac{\xi e^{-n\xi}}{1-e^{-n\xi}} = \frac{\xi e^{-\xi}}{1-e^{-\xi}}\left[\sum_1^\infty O\left(\frac{1}{m^2}\right)e^{-m\xi}-\log\left(1-e^{-\xi}\right)\right].\tag{9.09}
$$

For $\xi > 0$, the derived series on both sides converge absolutely and uniformly, and we may differentiate (9.09) term by term. On multiplication by ξ this yields us

$$
\begin{aligned}
&\xi\sum_1^\infty \Lambda(n)\,\frac{d}{dn\xi}\left(\frac{n\xi e^{-n\xi}}{e^{-n\xi}-1}\right)\\
&\quad= \frac{\xi e^{-\xi}(1-\xi-e^{-\xi})}{(1-e^{-\xi})^2}\left[\log(1-e^{-\xi})-\sum_1^\infty O\left(\frac{1}{m^2}\right)e^{-m\xi}\right]\\
&\qquad+\frac{\xi^2 e^{-\xi}}{1-e^{-\xi}}\left[\frac{e^{-\xi}}{1-e^{-\xi}}-\sum_1^\infty O\left(\frac{1}{m}\right)e^{-m\xi}\right]\\
&\quad= O(\xi)\,[O(\log\xi)-O(1)]+[\xi+O(\xi^2)]\left[\frac{1}{\xi}+O(\log\xi)\right]\\
&\quad= 1+O(\xi\log\xi)
\end{aligned}\tag{9.10}
$$

as $\xi\to 0$. If we put

$$
N_1(u) = \frac{d}{du}\left(\frac{ue^{-u}}{e^{-u}-1}\right)\tag{9.11}
$$

this leads us to

$$
\lim_{\xi\to 0}\xi\sum_1^\infty \Lambda(n)\,N_1(n\xi) = 1.\tag{9.12}
$$

This is a particular case of (6.14), which forms part of the hypothesis to Theorem XI‴. $\Lambda(n)$ is positive, as the hypothesis of that theorem further demands. The function $N_1(u)$ may be written

$$
N_1(u) = \frac{d}{du}\left(\frac{u}{1-e^u}\right) = \frac{1+(u-1)e^u}{(1-e^u)^2} = \begin{cases} O(1) & \text{as } u\to 0;\\ O(ue^{-u}) & \text{as } u\to\infty.\end{cases}\tag{9.13}
$$

Thus of that part of the hypothesis of Theorem XI‴ containing N_1, which is now both M and the whole class Σ, it only remains for us to verify the boundedness of

$$\xi \sum_1^\infty \Lambda(n)\, N_1(n\xi) \tag{9.14}$$

which follows from the fact that there is a K such that for $\xi > \xi_0 > 0$

$$\begin{aligned} \left| \xi \sum_1^\infty \Lambda(n)\, N_1(n\xi) \right| &\leqq K \sum_1^\infty n(n\xi e^{-n\xi}) \\ &= K\xi^2 \frac{d^2}{d\xi^2} \frac{e^{-\xi}}{1-e^{-\xi}} \to 0 \quad \text{as} \quad \xi \to 0 \end{aligned} \tag{9.15}$$

and from (9.12), and to verify further that

$$\int_0^\infty N_1(u)\, u^{ix}\, du \neq 0. \tag{9.16}$$

Now

$$\begin{aligned} &\int_0^\infty N_1(u)\, u^{ix}\, du \\ &= \int_0^\infty u^{ix}\, d\left(\frac{u}{1-e^u}\right) \\ &= \lim_{\lambda \to 0} \int_0^\infty u^{ix+\lambda}\, d\left(\frac{u}{1-e^u}\right) \\ &= -\lim_{\lambda \to 0} \int_0^\infty \frac{u}{1-e^u}\, d(u^{ix+\lambda}) \\ &= -\lim_{\lambda \to 0} (ix+\lambda) \int_0^\infty u^{ix+\lambda} \frac{du}{1-e^u} \qquad (9.17) \\ &= \lim_{\lambda \to 0} (ix+\lambda) \int_0^\infty u^{ix+\lambda} (e^{-u} + e^{-2u} + \cdots)\, du \\ &= \lim_{\lambda \to 0} (ix+\lambda)\, \Gamma(ix+\lambda+1)\, (1 + 2^{-(ix+\lambda+1)} + 3^{-(ix+\lambda+1)} + \ldots) \\ &= \lim_{\lambda \to 0} (ix+\lambda)\, \Gamma(ix+\lambda+1)\, \zeta(ix+\lambda+1) \\ &= ix\, \Gamma(ix+1)\, \zeta(ix+1). \end{aligned}$$

Let us take over from the theory of the Riemann zeta function the following facts:

(a) that the Riemann zeta function $\zeta(z)$ is analytic on the line with real part 1, except for a pole of the first order with principal part $1/(z-1)$;

(b) that the Riemann zeta function has no zeros with real part 1.

It then follows that

$$\int_0^\infty N_1(u)\, u^{ix}\, du \neq 0 \tag{9.18}$$

for any real x, and that

(9.19) $$\int_0^\infty N_1(u)\,du = \lim_{\lambda\to 0} \lambda\,\Gamma(\lambda+1)\,\zeta(\lambda+1) = 1.$$

Thus it follows from (6.35) that

(9.22) $$\lim_{N\to\infty} \frac{1}{N} \sum_{n=1}^{N} \Lambda(n) = 1.$$

Let us put $\pi(u)$ for the number of primes less than or equal to u, and let us write

(9.23) $$\varpi(u) = \pi(u) + \tfrac{1}{2}\pi(u^{1/2}) + \tfrac{1}{3}\pi(u^{1/3}) + \cdots.$$

It is easy to prove by elementary means that

(9.24) $$\lim_{u\to\infty} \frac{\pi(u)}{u} = 0.$$

Moreover,

(9.25) $$\begin{aligned} |\varpi(u) - \pi(u)| &= \left| \tfrac{1}{2}\pi(u^{1/2}) + \cdots + \frac{\pi\left(u^{\frac{1}{[\log u/\log 2+1]}}\right)}{[\log u/\log 2]+1} \right| \\ &\leqq u^{1/2}([\log u/\log 2]+1) \\ &= O(u^{1/2}\log u). \end{aligned}$$

Thus

(9.26) $$\lim_{u\to\infty} \frac{\varpi(u)}{u} = 0.$$

We may write (9.22) in the form

(9.27) $$\lim_{N\to\infty} \frac{1}{N} \int_0^N \log u\, d\varpi(u) = 1.$$

On integrating by parts, we see that

(9.28) $$\lim_{N\to\infty} \left\{ \frac{\varpi(N)\log N}{N} - \frac{1}{N}\int_0^N \frac{\varpi(u)\,du}{u} \right\} = 1.$$

Now

(9.29) $$\lim_{N\to\infty} \frac{1}{N} \int_0^N \frac{\varpi(u)}{u}\,du = 0$$

because of (9.26). Thus

(9.30) $$\lim_{N\to\infty} \frac{\varpi(N)\log N}{N} = 1$$

or

(9.31) $$\varpi(N) \sim \frac{N}{\log N}.$$

From (9.25) it follows that

(9.32) $$\pi(N) \sim \frac{N}{\log N}.$$

This is the famous prime number theorem of de la Valleé Poussin and Hadamard.

10. **Ikehara's Theorem.** The Landau Theorem (XVI) received several successive generalizations at the hands of Landau himself, and of Hardy and Littlewood, perhaps the most general of which was indicated by Hardy and Littlewood[36] to be the following:

THEOREM XVI. *Let:*

(i) *the series* $\Sigma a_n \lambda_n^{-s}$ *be absolutely convergent for* $\Re(s) > \sigma_0 > 0$;

(ii) *the function* $F(s)$ *defined by the series be regular for* $\Re(s) > c$ *where* $0 < c \leqq \sigma_0$ *and continuous for* $\Re(s) \geqq c$, *except for a simple pole with residue* g *at* $s = c$;

(iii)

$$F(s) = O(e^{C|t|}) \tag{10.01}$$

for some finite C, *uniformly for* $\sigma \geqq c$;

(iv)

$$\frac{\lambda_n}{\lambda_{n-1}} \to 1; \tag{10.02}$$

(v) a_n *be real, and satisfy one of the inequalities*

$$a_n > -K\lambda_n^{c-1}(\lambda_n - \lambda_{n-1}), \quad a_n < K\lambda_n^{c-1}(\lambda_n - \lambda_{n-1}) \tag{10.03}$$

or complex, and of the form

$$O\{\lambda_n^{c-1}(\lambda_n - \lambda_{n-1})\}. \tag{10.04}$$

Then

$$A_n = a_1 + a_2 + \cdots + a_n \sim \frac{g\lambda_n^c}{c}. \tag{10.05}$$

The vital change between the Landau and the Hardy-Littlewood theorem is the looser form of (iii), which replaces a restriction of the form

$$F(s) = O(|t|^k). \tag{9.04}$$

Both these restrictions are inessential, and the true theorem is that of Ikehara[37], which reads as follows:

THEOREM XVII. *Let* $\alpha(x)$ *be a monotone increasing function, and let*

$$\int_{1+0}^{\infty} x^{-u}\, d\alpha(x) = f(u) \tag{10.06}$$

[36] Hardy and Littlewood (8).

[37] Ikehara (1).

converge for $\Re(u) > 1$. *Let*

$$f(u) - \frac{A}{u-1} = g(u) \tag{10.07}$$

converge uniformly to a finite limit as

$$\Re(u) \to 1 \tag{10.08}$$

over any finite interval of the line $\Re(u) = 1$. *Then*

$$A = \lim_{N\to\infty} \frac{\alpha(N)}{N}. \tag{10.09}$$

To prove this, let us put

$$\beta(\xi) = \alpha(e^\xi)e^{-\xi} + \int_0^\xi e^{-\xi}\alpha(e^\xi)\,d\xi - A\xi. \quad (\xi > 0). \tag{10.10}$$

Thus

$$d\beta(\xi) = e^{-\xi}\,d\alpha(e^\xi) - A\,d\xi. \tag{10.11}$$

Let us assume—what is no essential restriction—that

$$\beta(x) = \beta(+0), \quad (-\infty \leqq x \leqq 0). \tag{10.12}$$

Then (10.07) becomes

$$g(u) = \int_{-\infty}^{\infty} e^{(1-u)\xi}\,d\beta(\xi), \quad [\Re(u) > 1]. \tag{10.13}$$

What we wish to prove is that

$$\lim_{\eta\to\infty} \int_{-\infty}^{\eta} e^{\xi-\eta}\,d\beta(\xi) = 0 \tag{10.14}$$

which is equivalent to (10.09).

If $\varepsilon > 0$ and η is real, we have

$$\begin{aligned}&\int_{-B}^{B}\left(1-\frac{|u|}{B}\right)g(iu+\varepsilon+1)\,e^{iu\eta}\,du \\ &\qquad = -\int_{-B}^{B}\left(1-\frac{|u|}{B}\right)du\int_{-\infty}^{\infty} e^{iu(\eta-\xi)}\,e^{-\varepsilon\xi}\,d\beta(\xi).\end{aligned} \tag{10.15}$$

As this double integral is absolutely convergent, it becomes

$$\begin{aligned}&\int_{-\infty}^{\infty} e^{-\varepsilon\xi}\,d\beta(\xi)\int_{-B}^{B}\left(1-\frac{|u|}{B}\right)e^{iu(\eta-\xi)}\,du \\ &\qquad = -\int_{-\infty}^{\infty}\frac{2(\cos B(\eta-\xi)-1)}{B(\eta-\xi)^2}\,e^{-\varepsilon\xi}\,d\beta(\xi).\end{aligned} \tag{10.16}$$

Now,

$$\begin{aligned}&\lim_{\varepsilon\to 0}\int_{-B}^{B}\left(1-\frac{|u|}{B}\right)g(iu+\varepsilon+1)\,e^{iu\eta}\,d\eta \\ &\qquad = \int_{-B}^{B}\left(1-\frac{|u|}{B}\right)g(iu+1)\,e^{iu\eta}\,d\eta.\end{aligned} \tag{10.17}$$

Thus

$$\lim_{\varepsilon \to 0} \int_{-\infty}^{\infty} \frac{2(\cos B(\eta - \xi) - 1)\, e^{-\varepsilon\xi}}{B(\eta - \xi)^2}\, d\beta(\xi) \tag{10.18}$$

exists. Remembering that

$$\lim_{\varepsilon \to 0} \int_{0}^{\infty} \frac{2(\cos B(\eta - \xi) - 1)}{B(\eta - \xi)^2}\, e^{-\varepsilon\xi}\, d\xi = \int_{0}^{\infty} \frac{2(\cos B(\eta - \xi) - 1)}{B(\eta - \xi)^2}\, d\xi, \tag{10.19}$$

and that

$$\frac{2(\cos B(\eta - \xi) - 1)}{B(\eta - \xi)^2}\, e^{(-1-\varepsilon)\xi}\, d\alpha(e^{\xi}) \tag{10.20}$$

is a non-positive integrand, we have, by a theorem due to Bray[38] and fundamental in the theory of the Stieltjes integral

$$\lim_{\varepsilon \to 0} \int_{0}^{\infty} \frac{2(\cos B(\eta - \xi) - 1)}{B(\eta - \xi)^2}\, e^{(-1-\varepsilon)\xi}\, d\alpha(e^{\xi}) = \int_{0}^{\infty} \frac{2(\cos B(\eta - \xi) - 1)}{B(\eta - \xi)^2}\, e^{-\xi}\, d\alpha(e^{\xi}) \tag{10.21}$$

and hence

$$\lim_{\varepsilon \to 0} \int_{-\infty}^{\infty} \frac{2(\cos B(\eta - \xi) - 1)}{B(\eta - \xi)^2}\, e^{-\varepsilon\xi}\, d\beta(\xi) = \int_{-\infty}^{\infty} \frac{2(\cos B(\eta - \xi) - 1)}{B(\eta - \xi)^2}\, d\beta(\xi). \tag{10.22}$$

Moreover,

$$-\lim_{\eta \to \infty} \int_{-\infty}^{\infty} \frac{2(\cos B(\eta - \xi) - 1)}{B(\eta - \xi)^2}\, d\beta(\xi) = \lim_{\eta \to \infty} \int_{-B}^{B} \left(1 - \frac{|u|}{B}\right) g(iu + 1)\, e^{iu\eta}\, du = 0 \tag{10.23}$$

because $\left(1 - \frac{|u|}{B}\right) g(iu + 1)$ is summable over $(-B, B)$. Similarly

$$\int_{-\infty}^{\infty} \frac{2(\cos B(\eta - \xi) - 1)}{B(\eta - \xi)^2}\, d\beta(\xi) \tag{10.24}$$

may be proved to be bounded because of (10.23) and (10.12).

We know that

$$\int_{n}^{n+1} d\beta(\xi) > -A \tag{10.25}$$

[38] Bray (1).

and that all the other conditions of Theorem XI are satisfied, with the possible exception of the non-vanishing of (6.15). To see that this is also satisfied, we need only reflect that

$$(10.26)\quad \int_{-\infty}^{\infty} \frac{2(\cos B\eta - 1)}{B\eta^2} e^{iu\eta}\, d\eta = -\frac{1}{2\pi}\left(1 - \frac{|u|}{B}\right) \qquad [|u| < B].$$

Thus there are no zeros common to all these functions for all values of B, and the non-vanishing of (6.15) follows.

It will be observed that the full force of our Tauberian method is scarcely needed for this theorem. In Theorem XI, which is the critical part of the proof, the difficulty of proof is considerably lessened if $K_1(x)$ assumes such a special form as $2(\cos Bx - 1)/Bx^2$.

As a corollary of Theorem XVII, we may prove:

THEOREM XVIII. *Let $\gamma(x)$ be a monotone increasing function, and let*

$$(10.27)\qquad \int_{1+0}^{\infty} x^{-u}\, d\gamma(x) = \varphi(u)$$

converge for $\Re(u) > 1$. *Let*

$$(10.28)\qquad F(u) = e^{\varphi(u)}(u-1)^A \qquad [0 < A < \tfrac{4}{3}]$$

when continued analytically, be regular for $\Re(u) = 1$, *and let it not vanish for* $u = 1$. *Then,*

$$(10.29)\qquad A = \lim_{N\to\infty} \frac{1}{N}\int_{1+0}^{N} \log x\, d\gamma(x).$$

To prove this, let us put

$$(10.30)\qquad \begin{cases} \displaystyle\int_{1+0}^{x} \log \xi\, d\gamma(\xi) = \alpha(x); \\ -\varphi'(u) = f(u). \end{cases}$$

The theorem then reduces itself to Theorem XVII, provided we can establish that (10.07) approaches a finite limit as $\Re(u) \to 1$. From the regularity of (10.28), it follows that

$$(10.31)\qquad \varphi(u) + A\log(u-1)$$

is regular for $\Re(u) = 1$, except for logarithmic singularities with *negative* infinities. It is also clear that there is no singularity for $u = 1$.

Now,

$$(10.32)\qquad \Re(\varphi(u)) = \int_{1+0}^{\infty} x^{-\Re(u)} \cos(\Im(u)\log x)\, d\gamma(x)$$

so that

$$(10.33)\qquad \Re(\varphi(1+\varepsilon+iv)) = \int_{1+0}^{\infty} x^{-(1+\varepsilon)} \cos(v\log x)\, d\gamma(x);$$

$$\Re(\varphi(1+\varepsilon+2iv)) = \int_{1+0}^{\infty} x^{-(1+\varepsilon)} \cos(2v \log x)\, d\gamma(x); \tag{10.34}$$

and

$$\varphi(1+\varepsilon) = \int_{1+0}^{\infty} x^{-(1+\varepsilon)}\, d\gamma(x). \tag{10.35}$$

Thus

$$\begin{aligned} &3\varphi(1+\varepsilon) + 4\Re(\varphi(1+\varepsilon+iv)) + \Re(\varphi(1+\varepsilon+2iv)) \\ &= \int_{1+0}^{\infty} x^{-(1+\varepsilon)} (3 + 4\cos v \log x + \cos(2v\log x))\, d\gamma(x) \end{aligned} \tag{10.36}$$

and since

$$\begin{aligned} 3 + 4\cos\varphi + \cos 2\varphi &= 3 + 4\cos\varphi + 2\cos^2\varphi - 1 \\ &= 2(1+\cos\varphi)^2 \\ &\geqq 0 \end{aligned} \tag{10.37}$$

it follows that

$$3\varphi(1+\varepsilon) + 4\Re(\varphi(1+\varepsilon+iv)) + \Re(\varphi(1+\varepsilon+2iv)) \geqq 0, \tag{10.38}$$

or

$$\Re(\varphi(1+\varepsilon+iv)) \geqq -\tfrac{3}{4}\varphi(1+\varepsilon) - \tfrac{1}{4}\Re(\varphi(1+\varepsilon+2iv)). \tag{10.39}$$

Thus

$$\begin{aligned} &\overline{\lim_{\varepsilon\to 0}} \frac{\Re(\varphi(1+\varepsilon+iv))}{\log\varepsilon} \\ &\leqq -\tfrac{3}{4}\lim_{\varepsilon\to 0}\frac{\varphi(1+\varepsilon)}{\log\varepsilon} - \tfrac{1}{4}\underline{\lim_{\varepsilon\to 0}}\frac{\Re(\varphi(1+\varepsilon+2iv))}{\log\varepsilon} \leqq \frac{3A}{4} < 1. \end{aligned} \tag{10.40}$$

On the other hand, if $1+iv$ is a logarithmic singularity of φ with a negative coefficient, it is a zero of $F(u)$ of integral order n, and

$$\overline{\lim_{\varepsilon\to 0}} \frac{\Re(\varphi(1+\varepsilon+iv))}{\log\varepsilon} = n > 1. \tag{10.41}$$

Thus $\varphi(u) + A\log(u-1)$ has no logarithmic singularities with negative coefficients, for $\Re(u) = 1$, and hence is analytic throughout this whole line. Thus, by differentiation,

$$f(u) - \frac{A}{u-1} \tag{10.42}$$

is analytic on the line in question, and the finiteness of the limit of (10.07) is established.

It will be observed that our proof, which completes the demonstration of Theorem XVIII, follows closely the lines of Landau's proof[39] that the Riemann zeta function has no zeros on the line $\Re(u) = 1$, and includes it

[39] Landau (2).

as a particular case. The prime number theorem itself is a particular case of Theorem XVIII. Let us put

$$\gamma(x) = \varpi(x). \tag{10.43}$$

Then

$$e^{\varphi(u)} = \zeta(u) \tag{10.44}$$

and the hypothesis of Theorem XVIII is manifestly satisfied. (10.29) then becomes

$$1 = \lim_{N\to\infty} \frac{1}{N}\int_{1+0}^{N} \log x \, d\varpi(x) \tag{9.27}$$

which we have shown to be equivalent to the prime number theorem.

Thus we have repeatedly shown that the prime number theorem is basically Tauberian in nature. It might consequently be expected that the more refined theorems as to the distribution of the primes, based on Riemann's hypothesis as to the distribution of the zeros of the zeta function, and established by Hardy, Littlewood and others, might be easy to establish on a Tauberian basis. Such a formula as

$$\lim_{\xi\to 0} \xi \sum_{1}^{\infty} (\Lambda(n) - 1)\, n^{\alpha}\, (n^{-\alpha}\, \xi^{-\alpha}\, N_1(n\xi)) = 0 \quad \left(0 < \alpha < \frac{1}{2}\right) \tag{10.45}$$

which follows from (9.10) much as does (9.12), appears to lend a certain color to this view. Here the theorem to which this seems to lead is

$$\lim_{N\to\infty} \frac{1}{N} \sum_{1}^{N} (\Lambda(n) - 1)\, n^{\alpha} = 0, \tag{10.46}$$

and the condition of non-vanishing on the Fourier transform becomes the Riemann condition

$$\zeta(ix + 1 - \alpha) \neq 0. \tag{10.47}$$

The author considers these hopes illusory and deceptive. Let it be noted that (10.45) does not form a satisfactory hypothesis to a Tauberian theorem until we have some hold on the boundedness of the mass distribution whose integral is

$$M(u) = \sum_{1}^{[u]} (\Lambda(n) - 1)\, n^{\alpha}. \tag{10.48}$$

Such information would already presuppose as much information as (10.45) can yield concerning all smaller values of α than occur in (10.46). In other words, Tauberian theorems merely transform a O into a o.

Another way of stating the same thing is to say that a Tauberian theorem always operates in the neighborhood of a single ordinate in the plane of the zeta function. This is because it depends on a division of the range of this function into near and remote parts, and because this division has validity in the theory of functions of a real variable, not in the theory of functions of a complex variable. On the other hand, the more refined properties of the distribution of primes depend on the behavior of the zeta function in the entire strip between ordinates $\frac{1}{2}$ and 1, inclusive, and can only be discussed with the aid of Cauchy's theorem.

Of course, no proof of the limitations of so vague a thing as a method has real mathematical cogency. At any time some super-Tauberian theorem may come to light and prove to be central in the utmost refinements of prime number theory. For the present, however, Tauberian theorems do not seem to lie on the main avenue of progress.

CHAPTER V.

SPECIAL APPLICATIONS OF TAUBERIAN THEOREMS.

11. **On the proof of special Tauberian theorems.** In the sequel, we shall show that the greater part of all known Tauberian theorems may be proved without great difficulty on the basis of the general theorems of the present paper. However, most of the particular theorems were proved in the first instance by entirely different methods. In individual cases, these methods are simpler and more direct than the general method here indicated. This is especially noticeable in the case of Karamata's[40] proof of the original Abel-Tauber theorem.

Being in possession of a general method, we may consider with advantage the particular methods and why they function. All Tauberian theorems of the type discussed in this paper are intimately related to the solution of an integral equation of the form

$$\int_{-\infty}^{\infty} F(y)\, G(x-y)\, dy = H(x). \tag{11.01}$$

The most direct and general method of solving such an equation is by the use of Fourier transforms. Nevertheless, there are many cases in which a repeated differentiation will reduce such an equation to a linear differential equation of finite order, and many more where the same repeated differentiation will lead to a differential equation of infinite order, but of manageable form. Thus it is appropriate in many cases to employ a technique

[40] Karamata (2), (3), (4).

of repeated differentiation, and this has been done by Hardy, Littlewood[41], and Vijayaraghavan,[42] though scarcely from an explicit consideration of the integral equations in question. So far, the successes of this method have been confined to cases where the analytic properties of the Fourier transform of F are extremely simple, and it has failed to throw any light on the Tauberian theorems of prime number theory.

The methods of Robert Schmidt[43] lie more along the lines of the present paper, in as much as he has seen the essential role played by the integral

$$\int_0^y K(x)\, x^u\, dx = R(u) \tag{11.02}$$

in the study of the kernel of a Tauberian theorem. However, he has devoted his attention to $R(u)$ for real integral arguments instead of for complex general arguments. Furthermore, Schmidt's general theorem concerns the unicity of the solution of his moment problem rather than its existence theory. As a consequence there is a wide gap between his general moment theorem and the particular Tauberian theorems which he obtains as corollaries. This gap he actually fills in in two cases, that of the Abel-Tauber theorem and that of the Borel-Tauber theorem, but he gives no general method by which it may be filled in in a new case. His actual procedure is closely allied to that of Hardy, Littlewood and Vijayaraghavan. Schmidt's chief service to the subject is in his great improvement of what may be called the auxiliary apparatus of the theory of Tauberian theorems, through his invention of the notions of "langsam abfallende Funktionen" and "gestrahlte Matrizen".

Karamata's elegant method leads to the study of the closure of a set of translations of a given function, and thus most closely approximates to that developed here. His function is

$$f(x) = e^x\, e^{-e^x} \tag{11.03}$$

and the problem is solved through Weierstrass' theory of polynomial approximation. The translations considered are accordingly those of the form

$$f(x + \log \mu). \tag{11.04}$$

Szász has carried further the study of this particular set of translations of a given function. This is a far more difficult study than that of the

[41] Cf. Littlewood (1).

[42] Vijayaraghavan (1), (2).

[43] Schmidt (1), (2).

closure of the complete set of translations, and here the general solution of the closure problem is not yet known to me.

12. Examples of kernels for which Tauberian theorems hold. Among the kernels admissible in the role of the $N(x)$ of Theorem XIII are:

(1) The Riesz kernels[44]

$$N(x) = \begin{cases} (1-x)^{\lambda}; & [0<x<1] \\ 0 \quad ; & [1<x] \end{cases} \quad [\lambda>0]. \tag{12.01}$$

(2) The Abel kernel

$$N(x) = e^{-x}; \tag{12.02}$$

(3) The kernel

$$N(x) = \frac{1}{1+x^2}, \tag{12.03}$$

corresponding to the method of summation of the series Σa_n, which gives as its partial sum the Abel average

$$\lambda \int_0^\infty f(x)\, e^{-\lambda x}\, dx \tag{12.04}$$

of the cosine series

$$f(x) = \sum a_n \cos nx. \tag{12.05}$$

In the respective cases we have

(1)

$$\int_0^\infty N(x)\, x^{z-1}\, dx = \frac{\Gamma(\lambda+1)\,\Gamma(z)}{\Gamma(\lambda+z+1)}; \tag{12.06}$$

(2)

$$\int_0^\infty N(x)\, x^{z-1}\, dx = \Gamma(z); \tag{12.07}$$

(3)

$$\int_0^\infty N(x)\, x^{z-1}\, dx = \frac{\pi}{2\sin\frac{\pi z}{2}}. \tag{12.08}$$

None of these functions vanishes for purely imaginary values of z.

A possible $N_1(\mu)$ of Theorem XI''' is

$$N_1(\mu) = \begin{cases} (\lambda+1)(1-\mu)^{\lambda}; & [0<\mu<1] \\ 0 & [1<\mu] \end{cases} \quad [\lambda \geqq 0]. \tag{12.09}$$

Here

$$\int_0^\infty N_1(\mu)\, \mu^{iu}\, d\mu = \frac{\Gamma(\lambda+2)\,\Gamma(iu+1)}{\Gamma(iu+\lambda+2)} \neq 0 \quad \text{if } u \text{ is real.} \tag{12.10}$$

[44] Riesz (1).

13. A theorem of Ramanujan. As a lemma in our further work, we shall find it convenient to introduce a theorem formulated by Ramanujan[45] and first proved by Hardy and Titchmarsh,[46] although with a formulation somewhat different from that here given. The theorem is intrinsically interesting, and is perhaps worth presenting in some detail.

Let $f(x)$ be a function of L_2, defined over $(0, \infty)$. Then $e^{x/2}f(e^x)$ will also belong to L_2 over $(-\infty, \infty)$; for

$$\int_{-\infty}^{\infty} |e^{x/2}f(e^x)|^2\,dx = \int_{-0}^{\infty} |f(x)|^2\,dx. \tag{13.01}$$

The Fourier transform of $e^{x/2}f(e^x)$ will be

$$\begin{aligned} h(u) &= \sqrt{\frac{1}{2\pi}}\ \underset{A\to\infty}{\text{l. i. m.}} \int_{-A}^{A} e^{iux}\,e^{x/2} f(e^x)\,dx \\ &= \frac{1}{\sqrt{2\pi}}\ \underset{\varepsilon\to 0}{\text{l. i. m.}} \int_{\varepsilon}^{1/\varepsilon} f(x)\,x^{iu-1/2}\,dx. \end{aligned} \tag{13.02}$$

The sine transform of $f(x)$ will be

$$g_1(y) = \sqrt{\frac{2}{\pi}}\ \underset{B\to\infty}{\text{l. i. m.}} \int_0^B f(x)\sin xy\,dx, \tag{13.03}$$

and its cosine transform,

$$g_2(y) = \sqrt{\frac{2}{\pi}}\ \underset{B\to\infty}{\text{l. i. m.}} \int_0^B f(x)\cos xy\,dx. \tag{13.04}$$

Let us put

$$k_1(u) = \frac{1}{\sqrt{2\pi}}\ \underset{\varepsilon\to 0}{\text{l. i. m.}} \int_{\varepsilon}^{1/\varepsilon} g_1(y)\,y^{iu-1/2}\,dy, \tag{13.05}$$

and

$$k_2(u) = \frac{1}{\sqrt{2\pi}}\ \underset{\varepsilon\to 0}{\text{l. i. m.}} \int_{\varepsilon}^{1/\varepsilon} g_2(y)\,y^{iu-1/2}\,dy. \tag{13.06}$$

We have

$$\begin{aligned} \frac{1}{z}\int_0^z g_1(y)\,dy &= \sqrt{\frac{2}{\pi}} \int_0^{\infty} f(x)\,dx\,\frac{1}{z}\int_0^z \sin xy\,dy \\ &= \sqrt{\frac{2}{\pi}} \int_0^{\infty} f(x)\,\frac{1-\cos xz}{xz}\,dx. \end{aligned} \tag{13.07}$$

Similarly

$$\frac{1}{z}\int_0^z g_2(y)\,dy = \sqrt{\frac{2}{\pi}} \int_0^{\infty} f(x)\,\frac{\sin xz}{xz}\,dx. \tag{13.08}$$

---

[45] Hardy and Littlewood (8).

[46] Hardy and Titchmarsh (1).

However, on an exponential transformation, these become

$$(13.09)\quad \int_{-\infty}^{\zeta} e^{\frac{\eta-\zeta}{2}} (e^{\eta/2} g_2(e^{\eta}))\, d\eta = \sqrt{\frac{2}{\pi}} \int_{-\infty}^{\infty} \sin(e^{\xi+\zeta})\, e^{-\frac{\xi+\zeta}{2}} (e^{\xi/2} f(e^{\xi}))\, d\xi,$$

and

$$(13.10)\quad \int_{-\infty}^{\zeta} e^{\frac{\eta-\zeta}{2}} (e^{\eta/2} g_1(e^{\eta}))\, d\eta = \sqrt{\frac{2}{\pi}} \int_{-\infty}^{\infty} (1-\cos(e^{\xi+\zeta}))\, e^{-\frac{\xi+\zeta}{2}} (e^{\xi/2} f(e^{\xi}))\, d\xi.$$

If we now make a Fourier transformation and make use of the Parseval theorem, it appears that

$$(13.11)\quad k_1(u) \int_0^{\infty} e^{-\eta/2} e^{iu\eta}\, d\eta = \sqrt{\frac{2}{\pi}}\, h(-u) \int_{-\infty}^{\infty} (1-\cos e^{\xi})\, e^{-\xi/2} e^{iu\xi}\, d\xi,$$

and

$$(13.12)\quad k_2(u) \int_0^{\infty} e^{-\eta/2} e^{iu\eta}\, d\eta = \sqrt{\frac{2}{\pi}}\, h(-u) \int_{-\infty}^{\infty} \sin e^{\xi}\, e^{-\xi/2} e^{iu\xi}\, d\xi.$$

Thus

$$(13.13)\quad \begin{aligned} \frac{k_1(u)}{h(-u)} &= \frac{\sqrt{\frac{2}{\pi}} \int_0^{\infty} (1-\cos x)\, x^{iu-3/2}\, dx}{\int_0^{\infty} e^{-\eta(1/2-iu)}\, d\eta} \\ &= \sqrt{\frac{2}{\pi}} \sin\left(iu+\frac{1}{2}\right)\frac{\pi}{2}\, \Gamma\left(iu+\frac{1}{2}\right). \end{aligned}$$

Here we make use of the formulae:

$$(13.131)\quad \int_0^{\infty} x^{\nu} \sin x\, dx = \Gamma(\nu+1) \cos\frac{\nu\pi}{2} \qquad [-1 \geqq \Re(\nu) > -2]$$

and

$$(13.132)\quad \int_0^{\infty} x^{\nu}(1-\cos x)\, dx = \Gamma(\nu+1) \sin\frac{\nu\pi}{2} \qquad [-1 > \Re(\nu) > -3].$$

Similarly

$$(13.14)\quad \frac{k_2(u)}{h(-u)} = \sqrt{\frac{2}{\pi}} \cos\left(iu+\frac{1}{2}\right)\frac{\pi}{2}\, \Gamma\left(iu+\frac{1}{2}\right).$$

The duality of the relation between $k_1(u)$ or $k_2(u)$ and $h(u)$ is shown by the familiar formulae of the gamma function,

$$(13.15)\quad \begin{aligned} &\frac{2}{\pi} \sin\left(iu+\frac{1}{2}\right)\frac{\pi}{2} \sin\left(-iu+\frac{1}{2}\right)\frac{\pi}{2}\, \Gamma\left(iu+\frac{1}{2}\right) \Gamma\left(-iu+\frac{1}{2}\right) \\ &= \frac{2}{\pi} \cos\left(iu+\frac{1}{2}\right)\frac{\pi}{2} \cos\left(-iu+\frac{1}{2}\right)\frac{\pi}{2}\, \Gamma\left(iu+\frac{1}{2}\right) \Gamma\left(-iu+\frac{1}{2}\right) \\ &= \frac{1}{\pi} \cos iu\pi\, \Gamma\left(iu+\frac{1}{2}\right) \Gamma\left(-iu+\frac{1}{2}\right) = 1. \end{aligned}$$

Thus

$$\left|\frac{k_2(u)}{h(-u)}\right| = 1, \qquad \left|\frac{k_1(u)}{h(-u)}\right| = 1, \tag{13.16}$$

and the real zeros of $k_1(u)$, $k_2(u)$, and $h(-u)$ are the same. In other words, if the integrals in question exist, the zeros of

$$\begin{cases} \dfrac{1}{\sqrt{2\pi}} \displaystyle\int_0^\infty f(x)\, x^{-iu-1/2}\, dx, \\ \dfrac{1}{\sqrt{2\pi}} \displaystyle\int_0^\infty g_1(y)\, y^{iu-1/2}\, dy, \\ \text{and } \dfrac{1}{\sqrt{2\pi}} \displaystyle\int_0^\infty g_2(y)\, y^{iu-1/2}\, dy, \end{cases} \tag{13.17}$$

are the same. This is a very valuable way of determining new kernels whose Fourier transforms (on a logarithmic scale) have no zeros.

14. **The summation of trigonometrical developments.** Let $f(x)$ be an even function of class L_2, and, as above, let

$$g_2(y) = \sqrt{\frac{2}{\pi}} \underset{B\to\infty}{\text{l. i. m.}} \int_0^B f(x) \cos xy\, dx. \tag{13.04}$$

Let $K(x)$ belong to L_2 over $(0, \infty)$, and let us form

$$x \int_0^\infty g_2(w)\, K(wx)\, dw. \tag{14.01}$$

This will equal

$$\frac{1}{x} \int_0^\infty z f(z) \frac{x}{z}\, k\left(\frac{z}{x}\right) dz, \tag{14.02}$$

where

$$k(z) = \sqrt{\frac{2}{\pi}} \underset{B\to\infty}{\text{l. i. m.}} \int_0^B K(w) \cos wz\, dw; \tag{14.03}$$

for the cosine transform of $xK(wx)$ is $k\left(\frac{z}{x}\right)$. Thus an average of $g_2(w)$ at infinity will appear formally as an average of $z f(z)$ about the origin, and an average of $g_2(w)$ about the origin will appear as an average of $z f(z)$ at infinity. The kernels in the two cases will be $K(x)$ and $\frac{1}{x} k(x)$. Now by § 13, we have formally

$$\left|\sqrt{\frac{1}{2\pi}} \int_0^\infty \frac{k(x)}{x} x^\lambda\, dx\right| = \left|\sqrt{\frac{1}{2\pi}} \int_0^\infty K(x)\, x^{-\lambda}\, dx\right|, \tag{14.04}$$

when $\Re(\lambda) = \frac{1}{2}$. If

$$(14.05) \qquad \sqrt{\frac{1}{2\pi}} \int_0^\infty \frac{k(x)}{x} x^\lambda \, dx \quad \text{and} \quad \sqrt{\frac{1}{2\pi}} \int_0^\infty K(x) x^{-\lambda} \, dx$$

converge for $0 < \Re(\lambda) \leq \frac{1}{2}$, it then follows that

$$(14.06) \qquad \sqrt{\frac{1}{2\pi}} \int_0^\infty \frac{k(x)}{x} x^{iu} \, dx \neq 0 \qquad [-\infty < u < \infty]$$

when, and only when,

$$(14.07) \qquad \frac{1}{\sqrt{2\pi}} \int_0^\infty K(x) x^{iu} \, dx \neq 0 \qquad [-\infty < u < \infty].$$

Thus if $k(x)/x$ in the role of $N_1(x)$, and $N_2(x)$ satisfy the other conditions specified in the hypothesis of Theorem XI, and (14.07) holds, it is possible to infer from

$$(14.08) \qquad \lim_{x \to {0 \atop \infty}} x \int_0^\infty g_2(w) K(wx) \, dw = A \int_0^\infty K(w) \, dw$$

to the conclusion that

$$(14.09) \qquad \lim_{x \to {0 \atop \infty}} x \int_0^\infty z f(z) N_2(zx) \, dz = A \int_0^\infty N_2(z) \, dz.$$

The condition that f belongs to L_2 may in many cases be very considerably altered and relaxed.

It is possible to read the relation between (14.01) and (14.02) in the reverse direction, and to infer from

$$(14.10) \qquad \lim_{x \to \infty} \int_0^\infty f(z) k\left(\frac{z}{x}\right) dz = A k(0)$$

and other conditions completing the hypothesis of Theorem XI, to

$$(14.11) \qquad \lim_{x \to 0} x \int_0^\infty g_2(w) K_2(wx) \, dw = A \int_0^\infty K_2(w) \, dw.$$

The condition that

$$(14.12) \qquad \frac{1}{\sqrt{2\pi}} \int_0^\infty k(x) x^{z-1} \, dx$$

should determine a function over a strip to the right of the origin, which, when continued has no zeros over the imaginary axis, becomes the same condition for

$$(14.13) \qquad \frac{1}{\sqrt{2\pi}} \int_0^\infty K(x) x^{-z} \, dx.$$

We thus arrive at a whole class of theorems relating the partial sum of a Fourier development with the average of the function represented about some point. For example, it is easy to prove:

THEOREM XVIII. *If $f(x)$ is a non-negative function of class L_2 over $(-\pi, \pi)$, if it is summable at any point by a Riesz mean of any positive order, or by an Abel mean, to any value A, it is so summable to A by Riesz means of all positive orders, and by an Abel mean. A necessary and sufficient condition that this should take place at a given point x is that*

$$\lim_{\varepsilon \to 0} \frac{1}{2\varepsilon} \int_{x-\varepsilon}^{x+\varepsilon} f(x)\, dx = A. \tag{14.14}$$

This theorem is due to Hardy and Littlewood.

15. **Young's criterion for the convergence of a Fourier series.** We now come to a region in which S. B. Littauer has done work, which constitutes the theorems proved in the present section. Young has proved the following theorem: *for the Fourier series of the integrable function $f(u)$ to converge to s for $u = x$, it is sufficient that*

$$\varphi(t) \to 0 \tag{15.01}$$

and that

$$\int_0^t |d(u\,\varphi(u))| = O(t) \tag{15.02}$$

for small t, where

$$\varphi(t) = \frac{1}{2}\{f(x+t) + f(x-t) - 2s\}. \tag{15.03}$$

Further work on this theorem has been done by Young, Pollard, Hardy and Littlewood. The chief theorem to which Hardy and Littlewood came was the following:

THEOREM XIX. *A necessary and sufficient condition that the Fourier series of the integrable function $f(u)$ be summable $(C, -1+\delta)$ for any positive δ is that*

$$\varphi_1(t) = o(t) \tag{15.04}$$

[*provided that*

$$\int_0^t |d(u\,\varphi(u))| = O(t), \qquad [0 \leqq t < \pi]] \tag{15.05}$$

where

$$\varphi_1(t) = \int_0^t \varphi(u)\, du. \tag{15.06}$$

This theorem in its original form is not directly adaptable to proof by the methods of this paper, inasmuch as Cesàro summation of fractional order does not depend on a kernel of the form $N(nx)$. On the other hand, if the Cesàro summation is replaced by Riesz summation, the theorem is reduced to a particular case of theorems already proved. We shall here content ourselves with proving this related theorem, as applied to Fourier integrals rather than series, in the case where $f(x)$ differs from zero only over a finite range.[47] Let us write

$$t\,\varphi(t) = \psi(t). \tag{15.07}$$

Then (15.05) becomes

$$\int_0^t |d\,\psi(t)| = O(t), \qquad [0 \leqq t < \pi] \tag{15.08}$$

which yields the boundedness of

$$\int_\mu^{2\mu} \lambda^{-1} |d\,\psi(\lambda)| - \int_\mu^{2\mu} \lambda^{-1}\, d\,\psi(\lambda). \tag{15.09}$$

Furthermore

$$\begin{aligned} \varphi_1(t) &= \int_0^t \varphi(u)\,du = \int_0^t \psi(u)\frac{du}{u} \\ &= \int_0^t \frac{du}{u}\int_0^u d\,\psi(v) = \int_0^t d\,\psi(v)\int_v^t \frac{du}{u} \\ &= \int_0^t \log\frac{t}{v}\, d\,\psi(v). \end{aligned} \tag{15.10}$$

Thus (15.04) becomes

$$\lim_{t\to 0}\frac{1}{t}\int_0^t \log\frac{t}{v}\, d\,\psi(v) = 0. \tag{15.11}$$

We have

$$\int_0^1 \log\lambda\ \lambda^{iu}\, d\lambda = -(iu+1)^{-2} \neq 0. \tag{15.12}$$

This is similar in form to (6.15).

The Riesz sum of kth order of the Fourier series of $f(x)-s$ for $x=0$ is

$$\begin{aligned} &\lim_{\varepsilon\to 0}\frac{1}{\varepsilon}\int_0^\infty \frac{\varphi(t)\,dt}{\Gamma(1+k)}\int_0^1 (1-\lambda)^k\cos\frac{\lambda t}{\varepsilon}\,d\lambda \\ &= \lim_{\varepsilon\to 0}\frac{1}{\varepsilon}\int_0^\infty \frac{dt}{t\,\Gamma(1+k)}\int_0^t d\psi(u)\int_0^1 (1-\lambda)^k\cos\frac{\lambda t}{\varepsilon}\,d\lambda \\ &= \lim_{\varepsilon\to 0}\frac{1}{\varepsilon}\int_0^\infty d\psi(u)\int_u^\infty \frac{dt}{t\,\Gamma(1+k)}\int_0^1 (1-\lambda)^k\cos\frac{\lambda t}{\varepsilon}\,d\lambda \\ &= \lim_{\varepsilon\to 0}\frac{1}{\varepsilon}\int_0^\infty d\psi(u)\int_{u/\varepsilon}^\infty \frac{dt}{t\,\Gamma(k+1)}\int_0^1 (1-\lambda)^k\cos\lambda t\,d\lambda. \end{aligned} \tag{15.13}$$

[47] It is possible to make the transition to Theorem XIX, but several of the steps require some consideration. In particular, the equivalence of Riesz and Cesàro summation for orders > -1 was proved by Riesz (Proc. Lond. Math. Soc., 2 (22), 412–419 (1924).

Here if $k > -1$,

$$\int_0^\infty \left| \int_0^1 (1-\lambda)^k \cos \lambda \tau \, d\lambda \right| d\tau \tag{15.14}$$

is finite, and the inversion of integration in (15.13) is hence applied to an absolutely convergent integral, in view of (15.08).

We have

$$\begin{aligned}
&\int_0^\infty \omega^{iu} \, d\omega \int_\omega^\infty \frac{d\tau}{\tau \Gamma(1+k)} \int_0^1 (1-\lambda)^k \cos \lambda \tau \, d\lambda \\
&= \frac{1}{iu+1} \int_0^\infty \frac{\omega^{iu}}{\Gamma(1+k)} \, d\omega \int_0^1 (1-\lambda)^k \cos \lambda \omega \, d\lambda \\
&= \frac{1}{iu+1} \int_0^\infty \frac{\omega^{iu}}{\Gamma(1+k)} \, d\omega \int_{\lambda=0}^{\lambda=1} (1-\lambda)^k \frac{d \sin \lambda \omega}{\omega} \\
&= \frac{1}{(iu+1)\Gamma(k)} \int_0^\infty \omega^{iu-1} \, d\omega \int_0^1 (1-\lambda)^{k-1} \sin \lambda \omega \, d\lambda \\
&= \frac{1}{(iu+1)\Gamma(k)} \int_0^1 (1-\lambda)^{k-1} \, d\lambda \int_0^\infty \omega^{iu-1} \sin \lambda \omega \, d\omega \\
&= \frac{1}{(iu+1)\Gamma(k)} \int_0^1 (1-\lambda)^{k-1} \lambda^{-iu} \, d\lambda \int_0^\infty z^{iu-1} \sin z \, dz \qquad (15.15) \\
&= \frac{1}{(iu+1)\Gamma(k)} \frac{\Gamma(k)\Gamma(1-iu)}{\Gamma(k+1-iu)} \Gamma(iu) \cos (iu-1) \frac{\pi}{2} \\
&= \frac{\Gamma(1-iu)\Gamma(iu)}{(iu+1)\Gamma(k+1-iu)} \sin \frac{\pi i u}{2} \\
&= \frac{\pi \sin \frac{\pi i u}{2}}{(iu+1)\Gamma(k+1-iu) \sin \pi i u} \\
&= \frac{\pi}{2(iu+1)\Gamma(k+1-iu) \cos \frac{\pi i u}{2}} \\
&\neq 0.
\end{aligned}$$

In proving this, we have made use of (13.131) and (13.132).

The other conditions of the hypothesis of Theorem XI′ are readily proved to be satisfied, and we have already shown that if (15.05) is satisfied, (15.11) *and*

$$\lim_{\varepsilon \to 0} \frac{1}{\varepsilon} \int_0^\infty \frac{\varphi(t)\, dt}{\Gamma(1+k)} \int_0^1 (1-\lambda)^k \cos \frac{\lambda t}{\varepsilon} \, d\lambda = 0, \tag{15.16}$$

are equivalent. Thus if (15.05) *is satisfied,* (15.04) *is completely equivalent to the statement that the Fourier series of $f(x)$ is summable to s by Riesz (or Cesàro) sums of any given order exceeding* -1.

16. **Tauberian theorems and asymptotic series.** Certain asymptotic problems arise in the discussion of the behavior of an integral

$$\int_a^b (g(t))^x \Phi(t)\,dt \tag{16.01}$$

for large values of x. By a change of variable, this problem may be reduced to the consideration of

$$\int_0^\infty e^{-xt} \Phi(t)\,dt. \tag{16.02}$$

In particular, let us discuss the situation which arises when $\nu > 0$ and

$$\lim_{x\to\infty} x^\nu \int_0^\infty e^{-xt} \Phi(t)\,dt = A, \tag{16.03}$$

which we may write

$$\lim_{x\to\infty} x \int_0^\infty (xt)^{\nu-1} e^{-xt} \frac{\Phi(t)}{t^{\nu-1}}\,dt = A. \tag{16.04}$$

If

$$x \int_0^\infty (xt)^{\nu-1} e^{-xt} \frac{\Phi(t)}{t^{\nu-1}}\,dt \tag{16.05}$$

is bounded and $\Phi(t)$ is positive, it will follow by Theorem XI″ that

$$\lim_{x\to\infty} x \int_0^{1/x} \frac{\Phi(t)}{t^{\nu-1}}\,dt = \frac{A}{\int_0^\infty t^{\nu-1} e^{-t}\,dt} = \frac{A}{\Gamma(\nu)}, \tag{16.06}$$

or that

$$\lim_{\varepsilon\to 0} \frac{1}{\varepsilon} \int_0^\varepsilon \frac{\Phi(t)}{t^{\nu-1}}\,dt = \frac{A}{\Gamma(\nu)}. \tag{16.07}$$

Again, it will follow that

$$\lim_{x\to\infty} x \int_{1/x}^\infty \frac{\Phi(t)}{t^{\nu-1}} \frac{dt}{x^2 t^2} = \frac{A}{\Gamma(\nu)}, \tag{16.08}$$

which yields

$$\lim_{\varepsilon\to 0} \varepsilon \int_0^\varepsilon t^{\nu-1} \Phi\left(\frac{1}{t}\right) dt = \frac{A}{\Gamma(\nu)}. \tag{16.09}$$

The condition that

$$\Gamma(iu+\nu) = \int_0^\infty e^{-t}\, t^{iu+\nu-1}\,dt \neq 0 \tag{16.10}$$

is obviously satisfied. Indeed the Tauberian theorems of this section (which are due to Ikehara[48]) differ only from those of the last section in that x tends to infinity instead of to 0.

If $\frac{\Phi(t)}{t^{\nu-1}}$ is bounded and tends to B at the origin and $\nu > 0$,

$$(16.11)\qquad \lim_{x\to\infty} x \int_0^\infty (xt)^{\nu-1} e^{-xt}\left[\frac{\Phi(t)}{t^{\nu-1}} - B\right] dt = 0.$$

Then

$$(16.12)\quad \lim_{x\to\infty} x \int_0^\infty (xt)^{\nu-1} e^{-xt} \frac{\Phi(t)}{t^{\nu-1}}\, dt = B x^{\nu} \int_0^\infty t^{\nu-1} e^{-xt}\, dt = B\Gamma(\nu).$$

Thus

$$(16.13)\qquad \lim_{x\to\infty} x^{\nu} \int_0^\infty e^{-xt}\, \Phi(t)\, dt = B\,\Gamma(\nu).$$

In particular, let there be a neighborhood of the origin in which $\Phi(t)$ is $\nu - 1$ times differentiable, with a bounded derivative of the $(\nu-1)$st order. Let

$$(16.14)\quad \Phi(0) = 0;\quad \Phi'(0) = 0;\quad \cdots\quad \Phi^{(\nu-2)}(0) = 0;\quad \Phi^{(\nu-1)}(0) = A$$

and let $\frac{\Phi(t)}{t^{\nu-1}}$ be bounded outside the neighborhood in question. Then

$$(16.03)\qquad \lim_{x\to\infty} x^{\nu} \int_0^\infty e^{-xt}\, \Phi(t)\, dt = A.$$

If now $\Phi(t) - A t^{\nu-1}$ has a bounded derivative of order $\mu - 1 > \nu - 1$ at the origin, and vanishes there with all its derivatives of order less than $\mu - 1$, and $\Psi(t)$ is a function which equals $\Phi(t) - \frac{A t^{\nu-1}}{\Gamma(\nu)}$ in some neighborhood of the origin, and for which $\frac{\Psi(t)}{t^{\mu-1}}$ is bounded and tends to A at the origin, we have

$$(16.15)\qquad \lim_{x\to\infty} x^{\mu} \int_0^\infty e^{-xt}\, \Psi(t)\, dt = A.$$

By a repetition of this process, it is easy to show that if

$$(16.16)\quad \Phi(t) = \frac{A_1 t^{\nu_1 - 1}}{\Gamma(\nu_1)} + \frac{A_2 t^{\nu_2 - 1}}{\Gamma(\nu_2)} + \cdots + \frac{A_k t^{\nu_k - 1}}{\Gamma(\nu_k)} + \Psi(t),$$

if $\Psi(t)$ is bounded and $(\mu - 1)$ times differentiable with bounded $(\mu-1)$st derivative in some neighborhood of the origin, if

[48] In a paper not yet published.

$$(16.17)\qquad \Psi(0) = 0; \quad \Psi'(0) = 0; \quad \cdots \quad \Psi^{(\mu-1)}(0) = 0,$$

and if $\mu > \nu_1,\ \mu > \nu_2,\ \cdots,\ \mu > \nu_k$, then

$$(16.18)\qquad \int_0^\infty e^{-xt}\,\Phi(t)\,dt = A_1 x^{-\nu_1} + A_2 x^{-\nu_2} + \cdots + A_k x^{-\nu_k} + O(x^{-\mu+\varepsilon})$$

as $x \to \infty$. This is an adequate basis for the theory of asymptotic expansions of Laplace integrals, and enables Wintner's work[49] on the subject to be greatly simplified.

CHAPTER VI.

KERNELS ALMOST OF THE CLOSED CYCLE.

17. **The reduction of kernels almost of the closed cycle to kernels of the closed cycle.** In the present section, we shall approach very close to the work of Robert Schmidt[50] in his discussion of "gestrahlte Mittelbildungen", although our terminology will be somewhat different. Up to this point, we have been discussing means of the form

$$(17.01)\qquad \int_{-\infty}^{\infty} K_1(\xi - x)\,f(\xi)\,d\xi$$

or

$$(17.02)\qquad \int_{-\infty}^{\infty} K_1(\xi - x)\,d\varphi(\xi),$$

or means only differing from these by a change of variable from ξ to a function of ξ. Let us now turn our attention to means of the form

$$(17.03)\qquad \int_{-\infty}^{\infty} K_1(\xi, x)\,d\varphi(\xi),$$

where

$$(17.04)\qquad K_1(\xi, x) = K_1(\xi - x) + K_1^*(\xi, x)$$

where $K_1(x)$ satisfies the condition that (5.02) converges and that (5.07) does not vanish, and where

$$(17.05)\qquad \lim_{x\to\infty} \sum_{-\infty}^{\infty} \max_{n \leq y \leq n+1} |K_1^*(y, x)| = 0.$$

Let $K_1(x, y)$ and $K_1(x)$ be continuous, and let

$$(17.06)\qquad \int_{-\infty}^{\infty} K_1(x, y)\,d\varphi(x) \quad \text{and} \quad \int_y^{y+1} |d\varphi(x)| - \varphi(y+1) + \varphi(y)$$

[49] Wintner (1).
[50] Schmidt (2).

be bounded. Let

$$\lim_{x \to \infty} \int_{-\infty}^{\infty} K_1(\xi, x)\, d\varphi(\xi) = A \int_{-\infty}^{\infty} K_1(\xi)\, d\xi. \tag{17.061}$$

The argument of Lemma XIa will need no substantial alteration to show that there is a Q such that

$$\int_y^{y+1} |d\varphi(x)| \leqq Q. \tag{6.04}$$

It will then follow from (17.05) that

$$\lim_{x \to \infty} \int_{-\infty}^{\infty} K_1(\xi - x)\, d\varphi(\xi) = A \int_{-\infty}^{\infty} K_1(\xi)\, d\xi \tag{5.03}$$

and that

$$\int_{-\infty}^{\infty} K_1(\xi - x)\, d\varphi(\xi)$$

is bounded. Hence, by Theorem X, if K_2 is any function belonging to N_1,

$$\lim_{x \to \infty} \int_{-\infty}^{\infty} K_2(\xi - x)\, d\varphi(\xi) = A \int_{-\infty}^{\infty} K_2(\xi)\, d\xi. \tag{5.04}$$

In Chapter III, the kernel $K(\xi - x)$ is replaced by a kernel of the form

$$M(\xi + \alpha - x) - M(\xi - x) \tag{17.07}$$

where $M(x)$ is a monotone function for which

$$M(-\infty) = 1, \qquad M(x) = O(e^{-ux}) \text{ at } +\infty; \qquad [\mu > 0] \tag{7.13}$$

and for which the function

$$\frac{1}{\sqrt{2\pi}} \int_{-\infty}^{\infty} M(x)\, e^{xz}\, dx \tag{7.05}$$

which is defined over a strip to the right of the origin, when continued analytically on to the imaginary axis, has no zeros there. We can replace the kernel $M(\xi - x)$ by a kernel of the form

$$M(\xi, x) = M_1(\xi - x) + M_2(\xi, x) \tag{17.08}$$

where $M_1(x)$ satisfies the conditions we have already laid down for $M(x)$, and where

$$\lim_{x \to \infty} \int_0^{\infty} |dM_2(\xi, x)| = 0 \tag{17.09}$$

and

$$M_2(\xi, x) = O(e^{-u\xi}) \text{ at } \infty \text{ uniformly.} \tag{17.10}$$

Under this change, Theorem XIV still remains valid.

18. **A Tauberian theorem of Hardy and Littlewood.**[51] Hardy and Littlewood have proved the following theorem:

THEOREM XX. *Let* $f(x) = \sum a_n x^n$ *be a power series with positive coefficients, and let*

$$f(x) \sim \frac{A}{(1-x)^\alpha} \quad (\alpha > 0,\ x \to 1). \tag{18.01}$$

Then

$$\sum_1^n a_k \sim \frac{n^\alpha A}{\Gamma(\alpha+1)} \quad (n \sim \infty). \tag{18.02}$$

We may write (18.01)

$$\sum a_n e^{-n\xi} \sim A\,\xi^{-\alpha} \tag{18.03}$$

or

$$\lim_{\xi\to 0} \xi \sum \frac{a_n}{n^{\alpha-1}} (n\xi)^{\alpha-1} e^{-n\xi} = A. \tag{18.04}$$

This falls under (6.30), and

$$\int_0^\infty \xi^{\alpha-1} e^{-\xi} \xi^{iu}\, d\xi = \Gamma(\alpha+iu) \neq 0. \tag{18.05}$$

Since $a_n/n^{\alpha-1} > 0$,

$$\lim_{\xi\to 0} \xi \sum_1^{[1/\xi]} \frac{a_n}{n^{\alpha-1}} (n\xi)^{\alpha-1} = A \frac{\int_0^1 \xi^{\alpha-1}\, d\xi}{\int_0^\infty \xi^{\alpha-1} e^{-\xi}\, d\xi} = \frac{A}{\Gamma(\alpha+1)} \tag{18.06}$$

because of Theorem XI'''. Writing m for $1/\xi$, and letting m become infinite through integral values, we get

$$\lim_{m\to\infty} \frac{1}{m^\alpha} \sum_1^m a_n = \frac{A}{\Gamma(\alpha+1)}, \tag{18.07}$$

which is only another way of writing (18.02).

This however is not the most general theorem proved by Hardy and Littlewood in this connection. They show that if

$$f(x) \sim \frac{A}{(1-x)^\alpha} L\left(\frac{1}{1-x}\right), \tag{18.08}$$

where

$$L(u) = (\log u)^{\alpha_1} (\log\log u)^{\alpha_2} \dots (\log^{(n)} u)^{\alpha_n} \tag{18.09}$$

and

$$u^\alpha L(u) \neq O(1) \text{ at } \infty, \tag{18.10}$$

then

$$\sum_1^n a_k \sim \frac{n^\alpha A\, L(n)}{\Gamma(\alpha+1)}. \tag{18.11}$$

[51] Hardy and Littlewood (4).

We may write our theorem to be proved in the form that if

$$\xi \sum_1^\infty \frac{a_n}{n^{\alpha-1}} (n\xi)^{\alpha-1} e^{-n\xi} \sim A L(\xi^{-1}), \tag{18.12}$$

then (18.11) follows. This we may again write

$$e^{-\mu} \sum_1^\infty \frac{a_n}{n^{\alpha-1}} e^{(\log n-\mu)(\alpha-1)} e^{-e^{\log n-\mu}} \sim A L(e^\mu) \tag{18.13}$$

as $\mu \to \infty$. Now, if $M(\xi)$ is a continuous function, defined over $(0, \infty)$, and asymptotic to $L(\xi)$ at ∞, we may show that:

$$\int_0^\infty e^{(u-\mu)\alpha} e^{-e^{(u-\mu)}} M(e^u)\, du \sim M(e^\mu) \int_{-\infty}^\infty e^u e^{-e^u} du \sim L(e^\mu). \tag{18.14}$$

Here $M(\xi)$ is introduced instead of $L(\xi)$ in order that we may have no trouble with finite singularities of $L(\xi)$.

To prove (18.14), let us reflect that

$$\begin{aligned} \int_0^\infty e^{(u-\mu)\alpha} e^{-e^{(u-\mu)}} M(e^u)\, du = &\int_0^{\mu-A} e^{(u-\mu)\alpha} e^{-e^{(u-\mu)}} M(e^u)\, du \\ &+\int_{\mu-A}^{\mu+A} e^{(u-\mu)\alpha} e^{-e^{(u-\mu)}} M(e^u)\, du \\ &+\int_{\mu+A}^\infty e^{(u-\mu)\alpha} e^{-e^{(u-\mu)}} M(e^u)\, du. \end{aligned} \tag{18.141}$$

Since $M(\xi)$ is asymptotic to $L(\xi)$, which is asymptotically increasing, the first integral is asymptotically less than

$$\begin{aligned} &L(e^u) \int_0^{\mu-A} e^{(u-\mu)\alpha} e^{-e^{(u-\mu)}} du \\ &\leqq L(e^\xi) \int_{-\infty}^{-A} e^u e^{-e^u} du \leqq L(e^\mu)\, e^{-A}. \end{aligned} \tag{18.142}$$

As to the second integral, it is asymptotically

$$L(e^\mu) \int_{-A}^A e^{\alpha u} e^{-e^u} du, \tag{18.143}$$

while the third one is ultimately less than

$$\begin{aligned} &e^{A^2} e^{-e^A} \int_{\mu+A}^\infty e^{-(u-\mu)} L(e^u)\, du \\ &\leqq e^{A^2} e^{-e^A} e^\mu \int_\mu^\infty L(e^u)\, d(-e^{-u}) \\ &= e^{A^2} e^{-e^A} \left[L(e^\mu) + e^\mu \int_\mu^\infty e^{-u} \frac{d}{du} (L(e^u))\, du\right] \\ &\leqq e^{A^2} e^{-e^A} \left[L(e^\mu) + C e^\mu \int_\mu^\infty e^{-u} u^{\alpha_1 - 1/2} du\right]. \end{aligned} \tag{18.144}$$

Now,

$$
(18.145)\quad
\begin{aligned}
\int_\mu^\infty e^{-u} u^{\alpha_1-1/2}\,du &= -\int_\mu^\infty u^{\alpha_1-1/2}\,d(e^{-u}) \\
&= e^{-\mu}\mu^{\alpha_1-1/2} + \int_\mu^\infty (\alpha_1-\tfrac{1}{2})\,e^{-u} u^{\alpha_1-3/2}\,du \\
&\leqq e^{-\mu}\left(\mu^{\alpha_1-1/2} + \int_\mu^\infty (\alpha_1-\tfrac{1}{2})\,u^{\alpha_1-3/2}\,du\right) \\
&\leqq 2e^{-\mu}\mu^{\alpha_1-1/2}.
\end{aligned}
$$

Thus by (18.144)

$$
(18.146)\quad \int_{\mu+A}^\infty e^{(u-\mu)\alpha} e^{-e^{(u-\mu)}} M(e^u)\,du \leqq e^{A^2} e^{-e^A}\left[L(e^\mu)+o(L(e^\mu))\right].
$$

Combining this with (18.141), (18.142) and (18.143), we see that

$$
(18.147)\quad
\begin{aligned}
\overline{\lim_{\mu\to\infty}}\left|\int_0^\infty e^{(u-\mu)\alpha} e^{-e^{(u-\mu)}} M(e^u)\,du - L(e^\mu)\int_{-\infty}^\infty e^u e^{-e^u}\,du\right| \Big/ L(e^\mu) \\
\leqq \varphi(A)
\end{aligned}
$$

where $\varphi(A)$ vanishes as A becomes infinite. Since A is arbitrarily large, (18.14) follows.

Consequently, by (18.14), if we put

$$
(18.15)\quad d\psi(u) = \frac{1}{u}\,d\sum_0^{[u]}\frac{a_n}{n^{\alpha-1}} - A\,M(u),
$$

(18.12) becomes

$$
(18.16)\quad \int_0^\infty (w\xi)^\alpha e^{-w\xi}\,d\psi(w)\to 0.
$$

On the other hand, we may write (18.11) in the form

$$
(18.17)\quad \int_0^{1/\xi} (w\xi)^\alpha\,d\psi(w)\to 0.
$$

Both integrals, (18.16) and (18.17), converge absolutely. We wish to make the transition from (18.16) to (18.17).

If $\alpha_1 \leqq 1$, $\psi(w)$ is "slowly decreasing" in the sense of Schmidt, we can apply Theorem XV, and it appears that $\psi(\lambda)\to 0$ as $\lambda\to\infty$. From this it follows that

$$
(18.18)\quad \int_0^{1/\xi}(w\xi)^\alpha\,d\psi(w) = \psi\left(\frac{1}{\xi}\right) - \psi(0) - \alpha\int_0^{1/\xi}\xi^\alpha w^{\alpha-1}\psi(w)\,dw \to 0.
$$

We now come to the more general case where $\alpha_1 > 1$. Our hypothesis becomes

$$
(18.19)\quad e^{-\mu}\sum\frac{a_n}{n^{\alpha-1}(\log n)^{[\alpha_1]}}\,e^{(\log u-\mu)(\alpha-1)}\,e^{-e^{\log n-\mu}}\,\frac{\log n^{[\alpha_1]}}{\mu}\sim A\,\varLambda(e^\mu),
$$

where

$$(18.20)\qquad \Lambda(u) = (\log u)^{\alpha_1-[\alpha_1]} (\log\log u)^{\alpha_2} \cdots (\log^{(n)} u)^{\alpha_n}.$$

The kernel of (18.19), in the sense of § 17, is

$$(18.21)\qquad \begin{aligned} e^{(\alpha-1)(y-x)}\, e^{-e^{y-x}} \left\{\frac{y}{x}\right\}^{[\alpha_1]} &= e^{(\alpha-1)(y-x)}\, e^{-e^{y-1}} \left\{1 - \frac{x-y}{x}\right\}^{[\alpha_1]} \\ &+ \left\{-[\alpha_1]\frac{x-y}{x} + \frac{[\alpha_1]\{[\alpha_1]-1\}}{2}\,\frac{(x-y)^2}{x^2} + \cdots + \frac{(x-y)}{x^{[\alpha_1]}}\right\}. \end{aligned}$$

It hence appears that this kernel is of the form indicated in (17.04) where K_1 and K_1^* satisfy the appropriate conditions. Thus if we put $M^*(\xi)\sim\Lambda(\xi)$ at ∞ and $M^*(\xi)$ is continuous, and if

$$(18.22)\qquad d\psi^*(u) = \frac{1}{u}\, d\sum_0^{[u]} \frac{a_n}{n^{\alpha-1}(\log n)^{[\alpha_1]}} - A\,M^*(u),$$

it follows just as in the case where $\alpha_1 \leqq 1$ that

$$(18.23)\qquad \int_0^{1/\xi} (w\,\xi)^{\alpha}\, d\psi^*(w) \to 0.$$

This however is only another way of writing (18.11).

19. The Tauberian theorem of Borel summation.[52] The Borel sum of the series with partial sums s_m is

$$(19.01)\qquad \lim_{x\to\infty} e^{-x} \sum_0^{\infty} \frac{s_n x^n}{n!} = s.$$

Let us put $n = [u^2]$ and

$$(19.02)\qquad \sum_1^n s_m (\sqrt{m} - \sqrt{m-1}) = f(u)$$

for $u > 0$. Let $f(u) = 0$ for negative u. Let us also put $x = y^2$. Then (19.01) becomes

$$(19.03)\qquad \lim_{y\to\infty} \int_0^{\infty} \frac{e^{-y^2}\, y^{2u^2}}{\Gamma(u^2+1)}\, \frac{d f(u)}{u - \sqrt{u^2-1}} = s.$$

Now

$$(19.04)\qquad \begin{aligned} \frac{e^{-y^2}\, y^{2u^2}}{\Gamma(u^2+1)\,(u-\sqrt{u^2-1})} &= \sqrt{\frac{2}{\pi}}\, e^{u^2-y^2} \left(\frac{y}{u}\right)^{2u^2} \left(1 + O\left(\frac{1}{u^2}\right)\right) \\ &= \sqrt{\frac{2}{\pi}}\, e^{-2(u-y)^2}\, e^{3u^2-4uy+y^2} \left(\frac{y}{u}\right)^{2u^2} \left(1 + O\left(\frac{1}{u^2}\right)\right). \end{aligned}$$

[52] For this section, cf. Schmidt (2); Vijayaraghavan (1).

Thus over the range $y-y^{1/6} \leqq u \leqq y+y^{1/6}$,

$$\frac{e^{-y^2}y^{2u^2}}{\Gamma(u^2+1)(u-\sqrt{u^2-1})} - \sqrt{\frac{2}{\pi}}\,e^{-2(u-y)^2} = \sqrt{\frac{2}{\pi}}\,e^{-2(u-y)^2}\,O\left(\frac{1}{y^2}\right) \tag{19.05}$$

because

$$\begin{aligned} & e^{3u^2-4uy+y^2}\left(\frac{y}{u}\right)^{2u^2} \\ &= \exp\left\{3u^2-4uy+y^2+2u^2\left[\frac{y-u}{u}-\frac{(y-u)^2}{2u^2}+O\left(\frac{(y-u)^3}{u^3}\right)\right]\right\} \\ &= 1+O\left(\frac{(y-u)^3}{u}\right) \\ &= 1+O(y^{-1/2}). \end{aligned} \tag{19.06}$$

Moreover, it is always true that

$$\begin{aligned} \frac{e^{-y^2}y^{2u^2}}{\Gamma(u^2+1)(u-\sqrt{u^2-1})} &= \sqrt{\frac{2}{\pi}}\,e^{u^2-y^2}\left(\frac{y}{u}\right)^{2u^2}\left(1+O\left(\frac{1}{u^2}\right)\right) \\ &\leqq \sqrt{\frac{2}{\pi}}\,e^{u^2-y^2}\,e^{\left(\frac{y-u}{u}\right)2u^2}\left(1+O\left(\frac{1}{u^2}\right)\right) \\ &= \sqrt{\frac{2}{\pi}}\,e^{-(u-y)^2}\left(1+O\left(\frac{1}{u^2}\right)\right). \end{aligned} \tag{19.07}$$

Thus the conditions (17.05) and (17.06) are satisfied if we put

$$\frac{e^{-y^2}y^{2u^2}}{\Gamma(u^2+1)(u-\sqrt{u^2-1})} = K_1(u,y) \tag{19.08}$$

and

$$\sqrt{\frac{2}{\pi}}\,e^{-2(u-y)^2} = K_1(u-y). \tag{19.09}$$

It follows at once that

$$\lim_{y\to\infty}\sqrt{\frac{2}{\pi}}\int_{-\infty}^{\infty} e^{-2(u-y)^2}\,df(u) = s \tag{19.10}$$

and that

$$\sqrt{\frac{2}{\pi}}\int_{-\infty}^{\infty} e^{-2(u-y)^2}\,df(u) \tag{19.101}$$

is bounded.

We now introduce again in its appropriate form Robert Schmidt's definition of a "langsam abfallende Folge". The sequence $\{s_m\}$ has this property if, whenever $q=q(p)$ $(p=0,1,\cdots)$ runs through such a sequence of indices that

$$q \geqq p \quad \text{and} \quad \frac{q-p}{\sqrt{p}} \to 0 \tag{19.11}$$

then

$$\varliminf_{p\to\infty} (s_q - s_p) \geqq 0. \tag{19.12}$$

Schmidt proves by elementary methods that if $s_0, s_1, \cdots$ is such a sequence, then

$$C(\Lambda) = \varlimsup_{\substack{p\to\infty \\ p\leqq q\leqq p+\Lambda\sqrt{p}}} (s_q - s_p) \tag{19.13}$$

exists, and

$$\lim_{\Lambda\to 0} C(\Lambda) = 0. \tag{19.14}$$

Furthermore, he shows that there exists a constant K, such that

$$s_q - s_p \geqq -K(\sqrt{q} - \sqrt{p}) \tag{19.15}$$

for all $p = 0, 1, \cdots$ and all $q \geqq p + \sqrt{p}$. Thus he shows that a K exists such that

$$\frac{s_p}{\sqrt{p+1}} \geqq -K \qquad (p = 0, 1, \cdots). \tag{19.16}$$

Let the sequence s_m be "langsam abfallend" and let (19.01) hold. Still following Schmidt, we see that

$$\begin{aligned} e^{-x}\sum_0^\infty \frac{s_n x^n}{n!} &= \sum_0^{[x]} \frac{e^{-x} s_n x^n}{n!} + \sum_{[x]+1}^{[x+\sqrt{x}]} \frac{e^{-x} s_n x^n}{n!} + \sum_{[x+\sqrt{x}]+1}^{\infty} \frac{e^{-x} s_n x^n}{n!} \\ &\geqq -K\sum_0^\infty \frac{e^{-x}\sqrt{n+1}\, x^n}{n!} + s_{[x]} \sum_{[x]+1}^{[x+\sqrt{x}]} \frac{e^{-x}x^n}{n!} - C(1)\sum_{[x]+1}^{[x+\sqrt{x}]} \frac{e^{-x}x^n}{n!}. \end{aligned} \tag{19.17}$$

If we build up the appropriate $f(u)$ as in (19.02), and make the transformation which led from (19.01) to (19.03), we get:

$$e^{-x}\sum_0^\infty \frac{\sqrt{n+1}\, x^n}{n!} \leqq \int_0^\infty \frac{e^{-x} x^{u^2}}{\Gamma(u^2+1)} \frac{d(u^2)}{u - \sqrt{u^2-1}}. \tag{19.18}$$

Again, by (19.07)

$$e^{-x}\sum_0^\infty \frac{\sqrt{n+1}\, x^n}{n!} \leqq \int_{-\infty}^\infty e^{-(u-\sqrt{x})^2} O(u)\, du = O(\sqrt{x}). \tag{19.19}$$

Obviously

$$\sum_{[x]+1}^{[x+\sqrt{x}]} \frac{e^{-x}x^n}{n!} < 1. \tag{19.20}$$

On the other hand

$$\lim_{x\to\infty} \sum_{[x]+1}^{[x+\sqrt{x}]} \frac{e^{-x}x^n}{n!} > \lim_{x\to\infty} \frac{e^{-x}x^x}{e^{-x-\sqrt{x}}\,(x+\sqrt{x}\,)^x \sqrt{2\pi(x+\sqrt{x}\,)}}\,[\sqrt{x}+1\,]$$

(19.21)
$$= \lim_{x\to\infty} \frac{e^{\sqrt{x}}}{\left(1+\dfrac{1}{\sqrt{x}}\right)^x \sqrt{2\pi}} = \left(\frac{e}{2\pi}\right)^{1/2}.$$

Thus (19.17) yields

(19.22) $$\overline{\lim_{x\to\infty}}\, x^{-1/2} s_{[x]} \leq \overline{\lim_{x\to\infty}} \sqrt{\frac{2\pi}{e}}\,[s x^{-1/2} + KO(1) + C(1)\,x^{-1/2}] = O(1),$$

which we may simplify and write

(19.23) $$s_{[x]} = O(x^{1/2}).$$

Let us now introduce the function

(19.24) $$g_\alpha(x) = \frac{1}{\alpha^2}\left[\int_{x+\alpha}^{x+2\alpha} - \int_x^{x+\alpha}\right] f(y)\,dy.$$

By (19.02),

(19.25) $$g_\alpha(x) = \frac{1}{\alpha^2}\left[\int_{(x+\alpha)^2}^{(x+2\alpha)^2} - \int_{x^2}^{(x+\alpha)^2}\right] \frac{dM}{2\sqrt{M}} \sum_1^{[M]} s_m(\sqrt{m}-\sqrt{m-1}\,).$$

Now,

$$\int_0^\nu \frac{dM}{2\sqrt{M}} \sum_1^{[M]} s_m(\sqrt{m}-\sqrt{m-1}\,)$$

(19.26)
$$= \sum_1^{[\nu]} s_m(\sqrt{m}-\sqrt{m-1}\,) \int_m^\nu \frac{dM}{2\sqrt{M}}$$

$$= \sum_1^{[\nu]} s_m(\sqrt{m}-\sqrt{m-1}\,)(\sqrt{\nu}-\sqrt{m}\,).$$

Substituting this in (19.25), and writing S for an average of the s_n's for $x^2 < n < (x+2\alpha)^2+1$, we have

$$g_\alpha(x) = \frac{1}{\alpha^2} \sum_1^{[(x+2\alpha)^2]} s_m(\sqrt{m}-\sqrt{m-1}\,)(x+2\alpha-\sqrt{m}\,)$$

$$- \frac{2}{\alpha^2} \sum_1^{[(x+\alpha)^2]} s_m(\sqrt{m}-\sqrt{m-1}\,)(x+\alpha-\sqrt{m}\,)$$

$$+ \frac{1}{\alpha^2} \sum_1^{[x^2]} s_m(\sqrt{m}-\sqrt{m-1}\,)(x-\sqrt{m}\,)$$

$$= \frac{1}{\alpha^2}\left\{ \sum_{[x^2]+1}^{[(x+\alpha)^2]} s_m(\sqrt{m}-\sqrt{m-1}\,)(\sqrt{m}-x) \right.$$

$$\left. + \sum_{[(x+\alpha)^2]+1}^{[(x+2\alpha)^2]} s_m(\sqrt{m}-\sqrt{m-1}\,)(x+2\alpha-\sqrt{m}\,) \right\}$$

$$(19.27)\quad = \frac{S}{\alpha^2}\left\{\sum_{[x^2]+1}^{[(x+\alpha)^2]}(\sqrt{m}-\sqrt{m-1})(\sqrt{m}-x) + \sum_{[(x+\alpha)^2]+1}^{[(x+2\alpha)^2]}(\sqrt{m}-\sqrt{m-1})(x+2\alpha-\sqrt{m})\right\}$$

$$= \frac{S}{\alpha^2}\left[\int_{x+\alpha}^{x+2\alpha} - \int_x^{x+\alpha}\right]\sqrt{[x^2]}\,dx$$

$$= S\left\{1+\frac{1}{\alpha^2}\left[\int_{x+\alpha}^{x+2\alpha} - \int_x^{x+\alpha}\right](\sqrt{[x^2]}-x)\,dx\right\}$$

$$= S\left\{1+O\left(\frac{1}{x^2}\right)+\frac{1}{\alpha^2}\left[\sum_{[x^2]}^{[(x+\alpha)^2]} - \sum_{[(x+\alpha)^2]}^{[(x+2\alpha)^2]}\right]\frac{(\sqrt{m+1}-\sqrt{m})^2}{2}\right\}$$

$$= S\left(1+O\left(\frac{1}{x^2}\right)\right).$$

It is indeed easy to show that as x increases, the weight of the later s_n's averaged in S increases at the expense of the weight of the earlier ones.

We may show that the condition that $\{s_m\}$ is "langsam abfallend" may be put in the form that if

$$(19.28)\qquad v \geqq u, \quad v-u\to 0$$

then

$$(19.29)\qquad \underline{\lim}_{u\to\infty}(s_{[v^2]}-s_{[u^2]}) \equiv 0.$$

Hence by (19.27) and (19.23), we may readily show that

$$(19.30)\qquad \underline{\lim}_{u\to\infty}(g_\alpha(v)-g_\alpha(u)) \geqq 0.$$

Again remembering (19.23), which makes all the integrals in question absolutely convergent, we have

$$\lim_{y\to\infty}\sqrt{\frac{2}{\pi}}\int_{-\infty}^{\infty}e^{-2(u-y)^2}g_\alpha(u)\,du$$

$$(19.31)\qquad = \lim_{y\to\infty}\sqrt{\frac{2}{\pi}}\int_{-\infty}^{\infty}e^{-2(u-y)^2}du\,\frac{1}{\alpha}\int_u^{u+\alpha}\frac{dv}{\alpha}\int_v^{v+\alpha}df(w)$$

$$= \lim_{y\to\infty}\frac{1}{\alpha^2}\sqrt{\frac{2}{\pi}}\int_y^{y+\alpha}dv\int_v^{v+\alpha}dw\int_{-\infty}^{\infty}e^{-2(u-w)^2}df(u)$$

$$= \lim_{y\to\infty}\sqrt{\frac{2}{\pi}}\int_{-\infty}^{\infty}e^{-2(u-y)^2}df(u) = s.$$

An integration by parts yields:

$$(19.32)\qquad s = \lim_{y\to\infty}\sqrt{\frac{2}{\pi}}\int_{-\infty}^{\infty}dg_\alpha(v)\int_v^{\infty}e^{-2(u-y)^2}du.$$

Similar methods establish the boundedness of

$$\sqrt{\frac{2}{\pi}} \int_{-\infty}^{\infty} d\, g_\alpha(v) \int_{v}^{\infty} e^{-2(u-y)^2}\, du. \tag{19.33}$$

We may then apply Theorem XIV, taking

$$M(x) = \sqrt{\frac{2}{\pi}} \int_{x}^{\infty} e^{-2u^2}\, du. \tag{19.34}$$

In this case

$$\varphi(z) = \sqrt{\frac{1}{2\pi z}}\, e^{z^2/8}. \tag{19.35}$$

Hence

$$\lim_{x\to\infty} g_\alpha(x) = s. \tag{19.36}$$

Let us now return to (19.27), making use of the existence of (19.13) and of (19.14). Then if x is sufficiently large,

$$S - s_{[x^2]} \geqq C(5\alpha) \tag{19.37}$$

and

$$s_{[(x+2\alpha)^2]} - S \geqq C(5\alpha). \tag{19.38}$$

It then follows from (19.27) and (19.36) that

$$\overline{\lim_{n\to\infty}}\, s_n \leqq s - C(5\alpha) \tag{19.39}$$

and

$$\underline{\lim_{n\to\infty}}\, s_n \geqq s + C(5\alpha). \tag{19.40}$$

Combining these, and remembering that $C(\Delta) \leqq 0$, we see that the following theorem holds:

THEOREM XXI. *Let* (19.01) *hold, and let* s_m *be "langsam abfallend". Then*

$$\lim_{n\to\infty} s_n = s. \tag{19.41}$$

CHAPTER VII.

A QUASI-TAUBERIAN THEOREM.

20. **The quasi-Tauberian theorem.** Up to the present point, all the Tauberian theorems we have discussed have involved some auxiliary condition of boundedness or positiveness or slow decrease. In the present chapter, we shall discuss a theorem without any such auxiliary condition as to the function averaged. The type of theorem is so fundamentally

different from that already discussed that the nomenclature, "Tauberian", seems to the author unfortunate. The theorems now to be discussed are in essence much closer to Abel's theorem than to Tauber's theorem.

By an integration by parts, we establish the following:

LEMMA XXIIa. *Let*

$$\int_B^C |F(x)|\,|df(x)| \tag{20.01}$$

exist for every $C > B$, and let

$$\int_B^\infty F(x)\,df(x) \tag{20.02}$$

exist as the limit of $\int_B^C F(x)\,df(x)$ as $C\to\infty$. Let

$$\int_B^C |G(x)|\,|df(x)| \tag{20.03}$$

exist for each $C > B$, let $G(x)$ and $F(x)$ be continuous, and let

$$\frac{G(x)}{F(x)}\to L \text{ as } x\to\infty; \qquad \int_D^\infty \left|d\,\frac{G(x)}{F(x)}\right| < \infty. \tag{20.04}$$

Then

$$\int_B^\infty G(x)\,df(x) \tag{20.05}$$

will exist, and

$$\begin{aligned}&\int_B^\infty G(x)\,df(x)\\ &= L\int_B^\infty F(x)\,df(x) - \int_B^\infty d\left(\frac{G(x)}{F(x)}\right)\int_B^x F(\xi)\,df(\xi).\end{aligned} \tag{20.06}$$

Now let us suppose that $f(x)$ is of limited total variation over every finite interval, that $K_1(x)$ is bounded and continuous, that

$$\int_{-\infty}^\infty |d(K_1(x)\,e^{-\lambda x})| < \text{const.}, \tag{20.061}$$

and that

$$\int_D^\infty K_1(-x)\,d\,f(x) \tag{20.07}$$

exists in the sense of (20.02). Let us suppose further that as $x\to-\infty$

$$K_1(x)\sim A_1 e^{\lambda x}, \qquad (\lambda>0) \tag{20.08}$$

where $A_1 \neq 0$. Then

(20.081)
$$\int_{-\infty}^{-D} \left| d\,\frac{e^{\lambda x}}{K_1(x)} \right| < \text{const.},$$
if D is sufficiently large that for $x \leqq -D$,

(20.082)
$$|e^{-\lambda x} K_1(x)| \geqq \text{const.} > 0.$$

As a particular case of (20.06), if B is large enough,

(20.083)
$$\int_B^\infty e^{-\lambda x}\, d\,f(x) = \frac{1}{A_1}\int_B^\infty K_1(-x)\,d\,f(x) - \int_B^\infty d\left(\frac{e^{-\lambda x}}{K_1(-x)}\right)\int_B^x K_1(-\xi)\,d\,f(\xi)$$
and hence

(20.084)
$$\int_B^\infty e^{-\lambda x}\, d\,f(x)$$
is bounded. A further application of (20.06) yields us

(20.085)
$$\int_B^\infty K_1(y-x)\,d\,f(x) = A_1 e^{\lambda y}\int_B^\infty e^{-\lambda x}\,d\,f(x) - \int_B^\infty d(K_1(y-x)\,e^{\lambda x})\int_B^x e^{-\lambda\xi}\,d\,f(\xi).$$
From this we may conclude that

(20.09)
$$\int_0^\infty K_1(y-x)\,d\,f(x) = \begin{cases} O(e^{\lambda y}) & \text{as } y \to \infty, \\ O(1) & \text{as } y \to -\infty \end{cases}$$
and that

(20.10)
$$\lim_{B\to\infty} e^{-\lambda y}\int_B^\infty K_1(y-x)\,d\,f(x) = 0$$
uniformly in y.

Now let $R(z)$ be a function for which

(20.11)
$$\int_{-\infty}^\infty e^{-\lambda z}\,|\,dR(z)|$$
and

(20.12)
$$\int_{-\infty}^\infty |\,dR(z)|$$
are finite. It follows from (20.10) that

(20.13)
$$\begin{aligned}\int_{-\infty}^\infty dR(y-z)\int_0^\infty K_1(z-x)\,d\,f(x) &= \lim_{B\to\infty}\int_{-\infty}^\infty dR(y-z)\int_0^B K_1(z-x)\,d\,f(x) \\ &= \lim_{B\to\infty}\int_0^B d\,f(x)\int_{-\infty}^\infty K_1(z-x)\,d\,R(y-z) \\ &= \int_0^\infty d\,f(x)\int_{-\infty}^\infty K_1(y-x-z)\,d\,R(z).\end{aligned}$$

Thus if

$$\lim_{y\to\infty}\int_0^\infty K_1(y-x)\,d\,f(x) = A\int_{-\infty}^\infty K_1(x)\,dx \tag{20.14}$$

then

$$\int_0^\infty K_1(y-x)\,d\,f(x) \tag{20.15}$$

will be bounded, by (20.09), and it follows that

$$\begin{aligned}
&\lim_{y\to\infty}\int_0^\infty d\,f(x)\int_{-\infty}^\infty K_1(y-x-z)\,d\,R(z)\\
&= \lim_{y\to\infty}\int_{-\infty}^\infty d\,R(y-z)\int_0^\infty K_1(z-x)\,d\,f(x)\\
&= \int_{-\infty}^\infty d\,R(-z)\lim_{y\to\infty}\int_0^\infty K_1(z+y-x)\,d\,f(x)\\
&= A\int_{-\infty}^\infty K_1(x)\,dx\int_{-\infty}^\infty d\,R(-z)\\
&= A\int_{-\infty}^\infty dx\int_{-\infty}^\infty K_1(x-z)\,d\,R(z).
\end{aligned} \tag{20.16}$$

Hence if $K_1(x)$ is bounded and continuous, if $f(x)$ is of limited total variation over every finite interval, if (20.14), (20.061) *and* (20.08) *hold, if* (20.11) *and* (20.12) *are finite, and if*

$$K_2(x) = \int_{-\infty}^\infty K_1(x-z)\,d\,R(z), \tag{20.17}$$

it follows that

$$\lim_{y\to\infty}\int_0^\infty K_2(y-x)\,d\,f(x) = A\int_{-\infty}^\infty K_2(x)\,dx. \tag{20.18}$$

This is a sufficiently important theorem to dignify by a number; we shall call it Theorem XXII.

A closely related proposition is:

THEOREM XXIII. *In the hypothesis of Theorem* XXII, *in case*

$$K_1(x) = 0, \qquad [x>0], \tag{20.19}$$

we may replace the assumption of the finiteness of (20.11) *and* (20.12) *by that of the finiteness of*

$$\int_{-\infty}^\infty |\,d\,R(z)\,|. \tag{20.20}$$

The conclusion remains valid.

To prove this, let us reflect that if (20.08) holds as $x \to \infty$, since

$$(20.21)\qquad \int_0^\infty K_1(y-x)\,df(x) = \int_y^\infty K_1(y-x)\,df(x),$$

it will follow from (20.06) that

$$(20.22)\qquad \begin{aligned}\int_y^\infty e^{-\lambda x}\,df(x) &= \frac{e^{-\lambda y}}{A_1}\int_y^\infty K_1(y-x)\,df(x)\\ &\quad -\int_y^\infty d\left(\frac{e^{-\lambda x}}{K_1(y-x)}\right)\int_y^x K_1(y-\xi)\,df(\xi)\\ &= e^{-\lambda y}\,O(1).\end{aligned}$$

By a precisely similar argument,

$$(20.23)\qquad \begin{aligned}&\int_B^\infty K_1(y-x)\,df(x)\\ &= A_1 e^{\lambda y}\int_B^\infty e^{-\lambda x}\,df(x) - \int_B^\infty d_x\,(e^{\lambda x}K_1(y-x))\int_B^\infty e^{-\lambda\xi}\,df(\xi)\\ &= A_1 e^{\lambda(y-B)}O(1) - \int_B^\infty d_x(e^{\lambda x}K_1(y-x))\,(e^{-\lambda B}O(1)+e^{-\lambda x}O(1))\\ &= O(e^{\lambda(y-B)}) \quad as \quad B\to\infty\end{aligned}$$

if $B>y$. Thus by (20.21), $\int_B^\infty K_1(y-x)\,df(x)$ is less than a function which is bounded and decreases monotonely to 0 as B becomes infinite. Consequently, by a theorem of Daniell on the Stieltjes integral,[53] in combination with the finiteness of (20.21),

$$(20.24)\qquad \lim_{B\to\infty}\int_{-\infty}^\infty dR(y-z)\int_B^\infty K_1(z-x)\,df(x) = 0.$$

Thus the inversions of integration in (20.13) are again permissible.

Formally and heuristically, (20.17) is equivalent to

$$(20.25)\qquad \frac{\int_{-\infty}^\infty K_2(x)\,e^{ux}\,dx}{\int_{-\infty}^\infty K_1(x)\,e^{ux}\,dx} = \int_{-\infty}^\infty e^{ux}\,dR(x).$$

If K_1 and K_2 are both $O(e^{-(\mu+\varepsilon)x})$ at $+\infty$ and $O(e^{+\varepsilon x})$ at $-\infty$,

$$(20.26)\qquad k_2(u) = \int_{-\infty}^\infty K_2(x)\,e^{ux}\,dx$$

and

$$(20.27)\qquad k_1(u) = \int_{-\infty}^\infty K_1(x)\,e^{ux}\,dx$$

[53] Daniell (1).

are both analytic over $-\varepsilon < \Re(u) < \varepsilon + \mu$. Formally,

$$R(x) = \frac{1}{2\pi i}\int_0^x d\xi \int_{-\infty}^{\infty} \frac{k_2(u)}{k_1(u)} e^{-u\xi}\, du, \tag{20.28}$$

the second integral being taken along any ordinate in the strip $-\varepsilon < \Re(u) < \varepsilon + \mu$.

Now let $K_1(x)$ belong to L_2. Let $k_2(u)/k_1(u)$ be analytic over $-\varepsilon \leqq \Re(u) \leqq \mu + \varepsilon$, and let it be quadratically summable over every ordinate in that strip. Then

$$e^{-\varepsilon x}\, \underset{A\to\infty}{\text{l.i.m.}} \int_{-A}^{A} \frac{k_2(iu)}{k_1(iu)} e^{-iux}\, du \tag{20.29}$$

and

$$e^{(\mu+\varepsilon)x}\, \underset{A\to\infty}{\text{l.i.m.}} \int_{-A}^{A} \frac{k_2(iu)}{k_1(iu)} e^{-iux}\, du \tag{20.30}$$

are quadratically summable. As a consequence,

$$\underset{A\to\infty}{\text{l.i.m.}} \int_{-A}^{A} \frac{k_2(iu)}{k_1(iu)} e^{-iux}\, du \tag{20.31}$$

and

$$e^{\mu x}\, \underset{A\to\infty}{\text{l.i.m.}} \int_{-A}^{A} \frac{k_2(iu)}{k_1(iu)} e^{-iux}\, du \tag{20.32}$$

are absolutely summable. If we assume R to be defined as in (20.28) and $\mu = \lambda$, we obtain:

THEOREM XXII′. *Let $K_1(x)$ be bounded and continuous. Let $f(x)$ be of limited total variation over every finite interval. Let* (20.14), (20.061) *and* (20.08) *hold. Let $k_2(u)$ and $k_1(u)$ be defined as in* (20.26) *and* (20.27), *respectively. Let $K_1(x)$ belong to L_2. Let $k_2(u)/k_1(u)$ be analytic over $-\varepsilon \leqq \Re(u) \leqq \lambda + \varepsilon$, and let it belong to L_2 over every ordinate in that strip. Then* (20.18) *follows.*

Similarly, we have:

THEOREM XXIII′. *In the hypothesis of Theorem* XXII′, *in case $K_1(x)$ vanishes for positive arguments, we may replace the strip $-\varepsilon \leqq \Re(u) \leqq \lambda + \varepsilon$ by the narrower strip $-\varepsilon \leqq \Re(u) \leqq \varepsilon$.*

21. Applications of the quasi-Tauberian theorem.

If

$$K^{(m)}(x) = \begin{cases} (m+1)(1-e^x)^m e^x; & [x<0] \\ 0 & [x>0] \end{cases} \tag{21.01}$$

and

$$^{(m)}K(x) = \frac{2e^x}{\pi}\int_0^1 (1-z)^m \cos(z e^x)\, dz; \qquad [m>0] \tag{21.02}$$

then (20.08) is clearly satisfied for $\lambda = 1$, and

$$K_1(x) = K^{(m)}(x) \tag{21.03}$$

or

$$K_1(x) = {}^{(m)}K(x). \tag{21.04}$$

If we put

$$\tau_m(u) = \int_{-\infty}^{\infty} K^{(m)}(x)\, e^{ux}\, dx = \frac{\Gamma(m+2)\,\Gamma(u+1)}{\Gamma(u+m+2)}, \tag{21.05}$$

and

$$\sigma_m(u) = \int_{-\infty}^{\infty} {}^{(m)}K(x)\, e^{ux}\, dx = \frac{\Gamma(m+1)}{\Gamma(m+1-u)\cos\frac{\pi u}{2}}, \tag{21.06}$$

then as $|\mathfrak{J}(u)| \to \infty$

$$\left|\frac{\sigma_m(u)}{\tau_n(u)}\right| \sim \frac{2\,\Gamma(m+1)}{\sqrt{2\pi}\,\Gamma(n+2)\, e^{\mathfrak{R}(u)-m}}\,|\mathfrak{J}(u)|^{\mathfrak{R}(u)-m+n+1/2}. \tag{21.07}$$

Thus if $n > m$, then

$$\lim_{y\to\infty} \int_0^{\infty} {}^{(m)}K(y-x)\, df(x) = A \int_0^{\infty} {}^{(m)}K(x)\, dx \tag{21.08}$$

will imply

$$\lim_{y\to\infty} \int_0^{\infty} K^{(n)}(y-x)\, df(x) = A \int_{-\infty}^{\infty} K^{(n)}(x)\, dx, \tag{21.09}$$

while if $m > n+1$, (21.09) will imply (21.08).

With a little manipulation, already indicated by the authors in question, this result is seen to be equivalent to an important theorem of Hardy and Littlewood,[54] which gives the necessary and sufficient condition for the summability of Fourier series and integrals by Cesàro sums of some order. In the integral form, the theorem reads as follows: *Let $f(x)$ be a measurable function defined over $(-\infty, \infty)$, and zero outside $(-A, A)$. Let*

$$\varphi(y) = \tfrac{1}{2}\{f(x+y) + f(x-y) - 2s\}. \tag{21.10}$$

Then if we write B_m for the proposition

$$\lim_{\lambda\to\infty} \lambda \lim_{\varepsilon\to 0} \int_{\varepsilon}^{1/\lambda} \varphi(y)\,(1-\lambda y)^{m-1}\, dy = 0, \tag{21.11}$$

and C_m for the proposition

$$\lim_{\lambda\to\infty} \lambda \lim_{\varepsilon\to 0} \int_{\varepsilon}^{\infty} \varphi(y)\, dy \int_0^1 (1-z)^m \cos\lambda y z\, dz = 0, \tag{21.12}$$

B_m implies $C_{m+\varepsilon}$ for $m \geqq 1$, while C_m implies $B_{m+1+\varepsilon}$ for $m > 0$.

[54] Hardy and Littlewood (12).

While the general method of the present section has been developed in an earlier paper of the author,[55] his final results were not stated correctly. The correct result is due to Bosanquet[56] and Paley, who have shown it to be a "best possible" result in both directions. Their theorem also applies to $m > -1$, and is otherwise somewhat more general.

CHAPTER VIII.

TAUBERIAN THEOREMS AND SPECTRA.

22. **A further type of asymptotic behavior.** The O and o symbols do not exhaust the possible terms in which we may describe the behavior of a function at infinity. A proposition which may be regarded as in some wise a generalization of

$$\lim_{x \to \infty} f(x) = A \tag{22.01}$$

is

$$\lim_{B \to \infty} \frac{1}{B} \int_0^B |f(x) - A|^2 \, dx = 0. \tag{22.02}$$

If $f(x)$ is bounded and measurable, (22.01) clearly implies (22.02), while (22.02) does not imply (22.01). Proposition (22.02) has a certain analogy to the different types of "strong convergence" to a limit which a function may exhibit: namely

$$\int_0^\infty |f(x) - A| \, dx \text{ converges} \tag{22.03}$$

and

$$\int_0^\infty |f(x) - A|^2 \, dx \text{ converges.} \tag{22.04}$$

The series analogues of the latter

$$\sum_1^\infty |s_n - A| \text{ converges} \tag{22.05}$$

and

$$\sum_1^\infty |s_n - A|^2 \text{ converges} \tag{22.06}$$

imply the ordinary convergence of s_n to A, but are not implied by it. On the other hand, neither (22.03) nor (22.04) is implied by ordinary convergence, although (22.04) implies (22.02). In contrast with propositions

[55] Wiener (5).

[56] Bosanquet (1), (2); Paley (1).

(22.03)—(22.06), which represent various types of "strong convergence", we shall express (22.02) in the usual language by saying that $f(x)$ is *strongly summable*[57] to A as $x \to \infty$. We shall say that A is a *sublimit* of $f(x)$, and shall write

$$\operatorname*{slm}_{x\to\infty} f(x) = A. \tag{22.07}$$

The sublimit of a function $f(x)$ differs markedly from the ordinary limit in that neither its existence nor its value are invariant if we replace x by a monotone function of x becoming infinite with x. On the other hand, the sublimit of $f(x)$ has a relation to the harmonic analysis of $f(x)$ far closer than does the ordinary limit.

Closely related to the notion of sublimit is that of subboundedness, which bears to the ordinary notion of boundedness much the same relation which that of sublimit does to the ordinary notion of limit. A function $f(x)$ is said to be *subbounded* if

$$\frac{1}{B}\int_0^B |f(x)|^2\,dx \tag{22.08}$$

is bounded. This juxtaposition of a notion of limit and a notion of boundedness suggests a generalized form of Tauberian theorem. To be specific, let us ask what conditions beyond the subboundedness of $f(x)$ are sufficient to make

$$\operatorname*{slm}_{x\to\infty} \int_0^\infty K_1(x-\xi) f(\xi)\,d\xi = A\int_{-\infty}^\infty K_1(\xi)\,d\xi \tag{22.09}$$

imply

$$\operatorname*{slm}_{x\to\infty} \int_0^\infty K_2(x-\xi) f(\xi)\,d\xi = A\int_{-\infty}^\infty K_2(\xi)\,d\xi. \tag{22.10}$$

This problem belongs to the range of ideas treated by the author in his work on generalized harmonic analysis. It gives a clearer picture of the real significance of Tauberian theorems. On the other hand, it does not at present offer an alternative approach to them, since Tauberian theorems of the type already discussed in this paper play an essential role in the establishment of a theory of generalized harmonic analysis. Results involving these theorems will be applied in this section.

To return to (22.09) and (22.10), no generality is lost by taking $A = 0$, as this simply amounts to replacing $f(x)$ by $f(x)-A$. Let us then take $A = 0$, and let us put $f(x) = 0$ for negative arguments, which

[57] Hardy and Littlewood (15).

is clearly permissible. We then wish to find a set of conditions which in conjunction with the boundedness for large B of

(22.11) $$\frac{1}{2B}\int_{-B}^{B}|f(x)|^2\,dx$$

(or what is the same, with the fact that $\frac{f(x)}{1+|x|}$ belongs to L_2) and the proposition

(22.12) $$\lim_{B\to\infty}\frac{1}{2B}\int_{-B}^{B}\left|\int_{-\infty}^{\infty}K_1(x-\xi)\,f(\xi)\,d\xi\right|^2 dx = 0$$

are sufficient to imply

(22.13) $$\lim_{B\to\infty}\frac{1}{2B}\int_{-B}^{B}\left|\int_{-\infty}^{\infty}K_2(x-\xi)\,f(\xi)\,d\xi\right|^2 d\xi = 0.$$

We shall prove the following:

THEOREM XXIV. *Let $f(x)$ be measurable, and let (22.11) be bounded. Let (22.12) be valid. Let K_1 and K_2 be measurable, and let*

(22.14) $$\int_{-\infty}^{\infty}(1+|x|)^{3+\varepsilon}\,|K_1(x)|^2\,dx<\infty$$ $$\left(\text{and hence}\int_{-\infty}^{\infty}(1+|x|)\,|K_1(x)|\,dx<\infty\right),$$

and

(22.15) $$\int_{-\infty}^{\infty}(1+|x|)^{3+\varepsilon}\,|K_2(x)|^2\,dx<\infty$$ $$\left(\text{and hence}\int_{-\infty}^{\infty}(1+|x|)\,|K_2(x)|\,dx<\infty\right).$$

Furthermore, let

(22.16) $$\int_{-\infty}^{\infty}K_1(x)\,e^{iux}\,dx \neq 0$$

for all real u. Then (22.13) is true.

For this theorem we shall need the following:

LEMMA XXIVa. *Let $f(x, u, \lambda)$ belong to L_2 as a function of u and be measurable in (x, u). Let $\int_{-\infty}^{\infty}|f(x, u, \lambda)|^2\,du$ be bounded, and let*

(22.17) $$\int_{-\infty}^{\infty}|f(x, u, \lambda)-f(x, u)|^2\,du\to 0$$

uniformly over any finite range of x as $\lambda\to\infty$. Let K be measurable, and let

$$\int_{-\infty}^{\infty} |K(x)|\, dx \tag{22.18}$$

be finite. Let

$$\int_{-\infty}^{\infty} K(x)\, f(x, u, \lambda)\, dx \tag{22.19}$$

exist for all λ. *Then*

$$\int_{-\infty}^{\infty} \left| \int_{-\infty}^{\infty} K(x) f(x, u, \lambda)\, dx - \int_{-\infty}^{\infty} K(x) f(x, u)\, dx \right|^2 du \to 0 \text{ as } \lambda \to \infty. \tag{22.20}$$

To prove this lemma, let us reflect that

$$\begin{aligned}
&\left[\int_A^B |K(x)|\, dx\right]^2 \max_x \int_{-\infty}^{\infty} |f(x, u, \lambda) - f(x, u)|^2\, du \\
&\geqq \int_A^B |K(x)|\, dx \int_A^B |K(y)|\, dy \int_{-\infty}^{\infty} du\, |f(x, u, \lambda) - f(x, u)| \\
&\qquad\qquad \times |f(y, u, \lambda) - f(y, u)| \\
&= \int_{-\infty}^{\infty} du \int_A^B |K(x)|\, dx \int_A^B |K(y)|\, dy\, |f(x, u, \lambda) - f(x, u)| \\
&\qquad\qquad \times |f(y, u, \lambda) - f(y, u)| \\
&\geqq \int_{-\infty}^{\infty} du \left| \int_A^B K(x)\, [f(x, u, \lambda) - f(x, u)]\, dx \right|^2.
\end{aligned} \tag{22.21}$$

By taking A large and B infinite, or $-B$ large and $-A$ infinite, we may make the left-hand member uniformly small. Over any finite range (A, B) the left-hand member tends to 0 as $\lambda \to \infty$. By combining these facts the lemma follows. It will be noted that it involves the existence almost everywhere of

$$\int_{-\infty}^{\infty} K(x)\, f(x, u)\, dx. \tag{22.22}$$

We may now return to the proof of our main theorem. In a previous paper[58], the author has shown that

$$s(u) = \frac{1}{2\pi} \int_{-1}^{1} f(x) \frac{e^{iux} - 1}{ix}\, dx + \underset{A\to\infty}{\text{l.i.m.}} \frac{1}{\pi} \left[\int_1^A + \int_A^{-1}\right] f(x) \frac{e^{iux}}{ix}\, dx \tag{22.23}$$

exists. Moreover,

$$\begin{aligned}
\int_{-\infty}^{\infty} |K_1(x-\xi) f(\xi)|\, d\xi &\leqq \int_{-\infty}^{\infty} |K_1(\xi)|\, |f(x-\xi)|\, d\xi \\
&\leqq \left[\int_{-\infty}^{\infty} (1+|\xi|)^2 |K_1(\xi)|^2\, d\xi \int_{-\infty}^{\infty} \frac{|f(x-\xi)|^2}{(1+|\xi|)^2}\, d\xi\right]^{1/2} \\
&< \infty,
\end{aligned} \tag{22.24}$$

[58] Wiener (7).

so that

$$
\begin{aligned}
&\frac{1}{2B}\int_{-B}^{B}\left|\int_{-\infty}^{\infty} K_1(x-\xi) f(\xi)\, d\xi\right|^2 dx \\
&\leqq \frac{1}{2B}\int_{-B}^{B} dx \int_{-\infty}^{\infty} |K_1(\xi)|\,|f(x-\xi)|\, d\xi \int_{-\infty}^{\infty} |K_1(\eta)|\,|f(x-\eta)|\, d\eta \\
&= \frac{1}{2B}\int_{-\infty}^{\infty} |K_1(\xi)|\, d\xi \int_{-\infty}^{\infty} |K_1(\eta)|\, d\eta \int_{-B}^{B} |f(x-\xi)|\,|f(x-\eta)|\, dx \qquad (22.25)\\
&\leqq \frac{1}{2B}\int_{-\infty}^{\infty} |K_1(\xi)|\, d\xi \int_{-\infty}^{\infty} |K_1(\eta)|\, d\eta \int_{-B-|\xi|-|\eta|}^{B+|\xi|+|\eta|} |f(x)|^2\, dx \\
&\leqq \frac{\text{const.}}{B}\int_{-\infty}^{\infty} |K_1(\xi)|\, d\xi \int_{-\infty}^{\infty} |K_1(\eta)|\, d\eta\, [B+|\xi|+|\eta|] \\
&< \text{const.}
\end{aligned}
$$

Thus the function

$$(22.26) \qquad f_1(x) = \int_{-\infty}^{\infty} K_1(x-\xi) f(\xi)\, d\xi$$

and the analogous function

$$(22.27) \qquad f_2(x) = \int_{-\infty}^{\infty} K_2(x-\xi) f(\xi)\, d\xi$$

define functions $s_1(u)$ and $s_2(u)$, corresponding to them in the same way in which $s(u)$ corresponds to $f(x)$.

In his previous paper,[58] the author has shown that

$$(22.28) \qquad s(u+\epsilon) - s(u-\epsilon) = \frac{1}{\pi} \underset{A\to\infty}{\text{l.i.m.}} \int_{-A}^{A} f(x) \frac{\sin \epsilon x}{x} e^{iux}\, dx.$$

Similarly

$$
\begin{aligned}
&s_1(u+\epsilon) - s_1(u-\epsilon) \\
(22.29) \qquad &= \frac{1}{\pi} \underset{A\to\infty}{\text{l.i.m.}} \int_{-A}^{A} \frac{\sin \epsilon x}{x} e^{iux}\, dx \int_{-\infty}^{\infty} K_1(\xi) f(x-\xi)\, d\xi.
\end{aligned}
$$

There is no trouble in modifying (22.28) into

$$(22.30) \qquad s(u+\epsilon) - s(u-\epsilon) = \frac{1}{\pi} \underset{A\to\infty}{\text{l.i.m.}} \int_{-A-\xi}^{A-\xi} f(x) \frac{\sin \epsilon x}{x} e^{iux}\, dx.$$

Thus

$$
\begin{aligned}
&s_1(u+\epsilon) - s_1(u-\epsilon) - (s(u+\epsilon) - s(u-\epsilon)) \int_{-\infty}^{\infty} K_1(\xi) e^{iu\xi}\, d\xi \\
&= \frac{1}{\pi}\Big[\underset{A\to\infty}{\text{l.i.m.}} \int_{-\infty}^{\infty} K_1(\xi)\, d\xi \int_{-A}^{A} f(x-\xi) \frac{\sin \epsilon x}{x} e^{iux}\, dx \\
(22.31) \qquad &\quad - \int_{-\infty}^{\infty} K_1(\xi)\, d\xi\, \underset{A\to\infty}{\text{l.i.m.}} \int_{-A}^{A} f(x-\xi) \frac{\sin \epsilon (x-\xi)}{x-\xi} e^{iux}\, dx \Big] \\
&= \frac{1}{\pi} \underset{A\to\infty}{\text{l.i.m.}} \int_{-\infty}^{\infty} K_1(\xi)\, d\xi \int_{-A}^{A} f(x-\xi) \left[\frac{\sin \epsilon x}{x} - \frac{\sin \epsilon (x-\xi)}{x-\xi}\right] e^{iux}\, dx.
\end{aligned}
$$

In the proof of this we have used our last lemma.

Clearly

$$\left| \frac{\sin \varepsilon x}{\varepsilon x} - \frac{\sin \varepsilon (x-\xi)}{\varepsilon (x-\xi)} \right| \leqq 2. \tag{22.32}$$

Again,

$$\begin{aligned} \left| \frac{\sin \varepsilon x}{\varepsilon x} - \frac{\sin \varepsilon (x-\xi)}{\varepsilon (x-\xi)} \right| &= \left| \int_{x-\xi}^{x} \frac{d}{dw} \left(\frac{\sin \varepsilon w}{\varepsilon w} \right) dw \right| \\ &= \left| \int_{x-\xi}^{x} \frac{\varepsilon^2 w \cos \varepsilon w - \varepsilon \sin \varepsilon w}{\varepsilon^2 w^2} dw \right| \\ &\leqq \varepsilon |\xi| \max_{w \text{ in} (x-\xi, x)} \left| \frac{\cos \varepsilon w}{\varepsilon w} - \frac{\sin \varepsilon w}{\varepsilon^2 w^2} \right| \qquad (22.321) \\ &\leqq 2\varepsilon |\xi| \max \frac{1}{|\varepsilon w|} \\ &\leqq \left| \frac{8\xi}{|x|+|\xi|} \right|. \end{aligned}$$

Thus and hence by an application of (22.32)

$$\frac{1}{\varepsilon \xi} \int_{-A}^{A} f(x-\xi) \left[\frac{\sin \varepsilon x}{x} - \frac{\sin \varepsilon (x-\xi)}{x-\xi} \right] e^{iux}\, dx \tag{22.33}$$

converges in the mean uniformly in u as $A \to \infty$ over any finite range of ξ. Hence by our lemma, we may write

$$\begin{aligned} &s_1(u+\varepsilon) - s_1(u-\varepsilon) - (s(u+\varepsilon) - s(u-\varepsilon)) \int_{-\infty}^{\infty} K_1(\xi)\, e^{iu\xi}\, d\xi \\ &= \frac{1}{\pi} \int_{-\infty}^{\infty} K_1(\xi)\, d\xi \operatorname*{l.i.m.}_{A\to\infty} \int_{-A}^{A} f(x-\xi) \left[\frac{\sin \varepsilon x}{x} - \frac{\sin \varepsilon (x-\xi)}{x-\xi} \right] e^{iux}\, dx. \end{aligned} \tag{22.34}$$

By a further use of (22.321) and our lemma, since

$$\operatorname*{l.i.m.}_{\varepsilon\to 0} \frac{1}{\varepsilon^{1/2}\xi} \operatorname*{l.i.m.}_{A\to\infty} \frac{1}{\pi} \int_{-A}^{A} f(x-\xi) \left[\frac{\sin \varepsilon x}{x} - \frac{\sin \varepsilon (x-\xi)}{x-\xi} \right] e^{iux}\, dx = 0 \tag{22.35}$$

uniformly in ξ, it follows that

$$\begin{aligned} \lim_{\varepsilon\to 0} \frac{1}{\varepsilon} \int_{-\infty}^{\infty} \Big| & s_1(u+\varepsilon) - s_1(u-\varepsilon) \\ & - (s(u+\varepsilon) - s(u-\varepsilon)) \int_{-\infty}^{\infty} K_1(\xi)\, e^{iu\xi}\, d\xi \Big|^2 du = 0. \end{aligned} \tag{22.36}$$

From (22.12) and a theorem of the previous paper,[59] it follows that

$$\lim_{\varepsilon\to 0} \frac{1}{\varepsilon} \int_{-\infty}^{\infty} |s_1(u+\varepsilon) - s_1(u-\varepsilon)|^2\, du = 0. \tag{22.37}$$

[59] Wiener (7), (5.52).

Combining (22.36) and (22.37), we get

$$(22.38)\quad \lim_{\varepsilon\to 0}\frac{1}{\varepsilon}\int_{-\infty}^{\infty}\left|\int_{-\infty}^{\infty}K_1(\xi)\,e^{iu\xi}\,d\xi\right|^2 |s(u+\varepsilon)-s(u-\varepsilon)|^2\,du = 0.$$

Since over any finite range, by (22.16),

$$(22.39)\quad \left|\int_{-\infty}^{\infty}K_1(\xi)\,e^{iu\xi}\,d\xi\right| > \text{const.},$$

it results that for any finite C,

$$(22.40)\quad \lim_{\varepsilon\to 0}\frac{1}{\varepsilon}\int_{-C}^{C}|s(u+\varepsilon)-s(u-\varepsilon)|^2\,du = 0.$$

As in (22.36), we have

$$(22.41)\quad \begin{aligned}\lim_{\varepsilon\to 0}\frac{1}{\varepsilon}\int_{-\infty}^{\infty}\Big|\, & s_2(u+\varepsilon)-s_2(u-\varepsilon)\\ & -(s(u+\varepsilon)-s(u-\varepsilon))\int_{-\infty}^{\infty}K_2(\xi)\,e^{iu\xi}\,d\xi\Big|^2 du = 0.\end{aligned}$$

This yields us

$$(22.42)\quad \begin{aligned}\lim_{\varepsilon\to 0}\frac{1}{\varepsilon}\int_{-C}^{C}\Big|\, & s_2(u+\varepsilon)-s_2(u-\varepsilon)\\ & -(s(u+\varepsilon)-s(u-\varepsilon))\int_{-\infty}^{\infty}K_2(\xi)\,e^{iu\xi}\,d\xi\Big|^2 du = 0.\end{aligned}$$

Combining this with (22.40), we get

$$(22.43)\quad \lim_{\varepsilon\to 0}\frac{1}{\varepsilon}\int_{-C}^{C}\left|s_2(u+\varepsilon)-s_2(u-\varepsilon)\right|^2 du = 0.$$

By our previous paper,[60]

$$(22.44)\quad \frac{1}{\pi\varepsilon}\int_{-\infty}^{\infty}|f_2(x)|^2\frac{\sin^2\varepsilon x}{x^2}\,dx = \frac{1}{2\varepsilon}\int_{-\infty}^{\infty}|s_2(u+\varepsilon)-s_2(u-\varepsilon)|^2\,du,$$

and

$$(22.45)\quad \begin{aligned}&\frac{1}{\pi\varepsilon}\int_{-\infty}^{\infty}\frac{\sin^2\varepsilon x}{x^2}\,dx\left|\frac{1}{2\lambda}\int_{-\lambda}^{\lambda}f_2(x+y)\,dy\right|^2\\ &\qquad = \frac{1}{2\varepsilon}\int_{-\infty}^{\infty}\frac{\sin^2\lambda u}{\lambda^2 u^2}|s_2(u+\varepsilon)-s_2(u-\varepsilon)|^2\,du.\end{aligned}$$

[60] Wiener (7), (5.52).

Now

$$\overline{\lim_{\varepsilon\to 0}}\left|\frac{1}{\pi\varepsilon}\int_{-\infty}^{\infty}|f_2(x)|^2\frac{\sin^2\varepsilon x}{x^2}dx-\frac{1}{\pi\varepsilon}\int_{-\infty}^{\infty}\frac{\sin^2\varepsilon x}{x^2}dx\left|\frac{1}{2\lambda}\int_{-\lambda}^{\lambda}f_2(x+y)\,dy\right|^2\right|$$

$$\leqq \overline{\lim_{\varepsilon\to 0}}\frac{\text{const.}}{\varepsilon}\int_{-\infty}^{\infty}\left|f_2(x)-\frac{1}{2\lambda}\int_{-\lambda}^{\lambda}f_2(x+y)\,dy\right|^2\frac{\sin^2\varepsilon x}{x^2}dx$$

$$=\overline{\lim_{\varepsilon\to 0}}\frac{\text{const.}}{\varepsilon}\int_{-\infty}^{\infty}\left|\int_{-\infty}^{\infty}f(x-\xi)\right.$$

$$\left.\times\left[K_2(\xi)-\frac{1}{2\lambda}\int_{-\lambda}^{\lambda}K_2(\xi+y)\,dy\right]d\xi\right|^2\frac{\sin^2\varepsilon x}{x^2}dx$$

$$\leqq \overline{\lim_{B\to\infty}}\ \text{const.}\frac{1}{2B}\int_{-B}^{B}\left|\int_{-\infty}^{\infty}f(x-\xi)\right. \tag{22.46}$$

$$\left.\times\left[K_2(\xi)-\frac{1}{2\lambda}\int_{-\lambda}^{\lambda}K_2(\xi+y)\,dy\right]d\xi\right|^2dx$$

$$\leqq \text{const.}\ \overline{\lim_{B\to\infty}}\frac{1}{B}\int_{-\infty}^{\infty}\left|K_2(\xi)-\frac{1}{2\lambda}\int_{-\lambda}^{\lambda}K_2(\xi+y)\,dy\right|d\xi$$

$$\times\int_{-\infty}^{\infty}\left|K_2(\eta)-\frac{1}{2\lambda}\int_{-\lambda}^{\lambda}K_2(\eta+z)\,dz\right|d\eta\,(B+|\xi|+|\eta|)$$

$$\leqq \text{const.}\ \overline{\lim_{B\to\infty}}\left[\int_{-\infty}^{\infty}\left|K_2(\xi)-\frac{1}{2\lambda}\int_{-\lambda}^{\lambda}K_2(\xi+y)\,dy\right|\left(1+\frac{|\xi|}{B}\right)d\xi\right]^2,$$

where

$$\overline{\lim_{B\to\infty}}\int_{-\infty}^{\infty}\left|K_2(\xi)-\frac{1}{2\lambda}\int_{-\lambda}^{\lambda}K_2(\xi+y)\,dy\right|\left(1+\frac{|\xi|}{B}\right)d\xi$$
$$=\int_{-\infty}^{\infty}\left|K_2(\xi)-\frac{1}{2\lambda}\int_{-\lambda}^{\lambda}K_2(\xi+y)\,dy\right|d\xi. \tag{22.47}$$

Hence

$$\overline{\lim_{\varepsilon\to 0}}\left|\frac{1}{\pi\varepsilon}\int_{-\infty}^{\infty}|f_2(x)|^2\frac{\sin^2\varepsilon x}{x^2}dx-\frac{1}{\pi\varepsilon}\int_{-\infty}^{\infty}\frac{\sin^2\varepsilon x}{x^2}dx\left|\frac{1}{2\lambda}\int_{-\lambda}^{\lambda}f_2(x+y)\,dy\right|^2\right|$$

$$\leqq \text{const.}\left[\int_{-\infty}^{\infty}\left|K_2(\xi)-\frac{1}{2\lambda}\int_{-\lambda}^{\lambda}K_2(\xi+y)\,dy\right|d\xi\right]^2. \tag{22.48}$$

As we have seen in section 1,

$$\lim_{\lambda\to 0}\int_{-\infty}^{\infty}\left|K_2(\xi)-\frac{1}{2\lambda}\int_{-\lambda}^{\lambda}K_2(\xi+y)\,dy\right|d\xi=0, \tag{22.49}$$

from which it immediately follows that

$$\lim_{\lambda\to 0}\overline{\lim_{\varepsilon\to 0}}\left|\frac{1}{\pi\varepsilon}\int_{-\infty}^{\infty}|f_2(x)|^2\frac{\sin^2\varepsilon x}{x^2}\,dx - \frac{1}{\pi\varepsilon}\int_{-\infty}^{\infty}\frac{\sin^2\varepsilon x}{x^2}\,dx\left|\frac{1}{2\lambda}\int_{-\lambda}^{\lambda}f_2(x+y)\,dy\right|^2\right| = 0. \tag{22.50}$$

In combination with (22.44) and (22.45) this yields us

$$\lim_{\lambda\to 0}\overline{\lim_{\varepsilon\to 0}}\frac{1}{2\varepsilon}\int_{-\infty}^{\infty}\left(1-\frac{\sin^2\lambda u}{\lambda^2u^2}\right)|s_2(u+\varepsilon)-s_2(u-\varepsilon)|^2\,du = 0. \tag{22.51}$$

Since

$$1-\frac{\sin^2\lambda u}{\lambda^2 u^2} > \frac{1}{2} \quad \text{if} \quad u > \frac{2}{\lambda} \tag{22.52}$$

it results at once that

$$\lim_{C\to\infty}\overline{\lim_{\varepsilon\to 0}}\frac{1}{2\varepsilon}\left[\int_C^{\infty}+\int_{-\infty}^{-C}\right]|s_2(u+\varepsilon)-s_2(u-\varepsilon)|^2\,du = 0. \tag{22.53}$$

Combining this with (22.43), we may readily conclude that

$$\lim_{\varepsilon\to 0}\frac{1}{2\varepsilon}\int_{-\infty}^{\infty}|s_2(u+\varepsilon)-s_2(u-\varepsilon)|^2\,du = 0, \tag{22.54}$$

or in view of a theorem of our previous paper, that

$$\lim_{T\to\infty}\frac{1}{2T}\int_{-T}^{T}|f_2(x)|^2\,dx = 0. \tag{22.13}$$

This establishes Theorem XXIV. It will be noted that this theorem has an immediate extension in the direction of Theorem XI′.

There is another Tauberian theorem concerning strong summability which we may discuss here. It is due to Hardy and Littlewood.[61] The particular case of it which most directly interests us is the following:

THEOREM XXV. *Let*

$$\frac{1}{\xi}\int_0^{\xi}|f(x)|^2\,dx \tag{22.55}$$

be bounded, and let

$$\lim_{\varepsilon\to 0}\frac{1}{\varepsilon}\int_0^{\varepsilon}f(x)\,dx = 0. \tag{22.56}$$

[61] Hardy and Littlewood (16).

Then if

$$(22.57)\qquad s_n(0) = \frac{1}{\pi}\int_0^\pi f(x)\frac{\sin\frac{2n+1}{2}x}{\sin\frac{x}{2}}\,dx$$

it follows that

$$(22.58)\qquad \lim_{N\to\infty}\frac{1}{N}\sum_0^N |s_n(0)|^2 = 0.$$

In the proof of this theorem, it is easy to see that we may replace the conclusion (22.58) which we are to establish, by the equivalent

$$(22.59)\qquad \lim_{H\to\infty}\frac{1}{H}\int_0^H\left|\frac{2}{\pi}\int_0^\pi f(x)\frac{\sin ux}{x}\,dx\right|^2 du = 0.$$

By a Tauberian theorem of the type of Theorem XI, already established in a previous paper of the author,[58] it follows that (22.59) is completely equivalent to

$$(22.60)\qquad \lim_{\varepsilon\to 0}\frac{2}{\pi}\int_0^\infty\frac{\sin^2\varepsilon u}{\varepsilon u^2}\left|\frac{2}{\pi}\int_0^\pi f(x)\frac{\sin ux}{x}\,dx\right|^2 du.$$

On the assumption that $f(x) = 0$ if $x > \pi$, we have by the Plancherel theorem

$$\frac{2}{\pi}\int_0^\infty\frac{\sin^2\varepsilon u}{\varepsilon u^2}\left|\frac{2}{\pi}\int_0^\pi f(x)\frac{\sin ux}{x}\,dx\right|^2 du$$

$$= \frac{2}{\pi\varepsilon}\int_0^\infty\left|\frac{2}{\pi}\int_0^\pi f(x)\frac{\sin\varepsilon u\sin ux}{ux}\,dx\right|^2 du$$

$$= \frac{2}{\pi\varepsilon}\int_0^\infty\left|\frac{2}{\pi}\int_0^\pi\frac{\sin\varepsilon u\sin ux}{u}\,d\int_x^\infty\frac{f(\xi)}{\xi}\,d\xi\right|^2 du$$

$$= \frac{2}{\pi\varepsilon}\int_0^\infty\left|\frac{2}{\pi}\int_0^\infty\sin\varepsilon u\cos ux\,dx\int_x^\infty\frac{f(\xi)}{\xi}\,d\xi\right|^2 du$$

$$(22.61)\qquad = \frac{2}{\pi\varepsilon}\int_0^\infty\left|\frac{1}{\pi}\int_0^\infty(\sin u(x+\varepsilon)-\sin u(x-\varepsilon))\,dx\int_x^\infty\frac{f(\xi)}{\xi}\,d\xi\right|^2 du$$

$$= \frac{2}{\pi\varepsilon}\int_0^\infty\left|\frac{1}{2\pi}\int_{-\infty}^\infty(\sin u(x+\varepsilon)-\sin u(x-\varepsilon))\,dx\int_{|x|}^\infty\frac{f(\xi)}{\xi}\,d\xi\right|^2 du$$

$$= \frac{2}{\pi\varepsilon}\int_0^\infty\left|\frac{1}{2\pi}\int_{-\infty}^\infty\sin ux\,dx\int_{|x-\varepsilon|}^{|x+\varepsilon|}\frac{f(\xi)}{\xi}\,d\xi\right|^2 du$$

$$= \frac{1}{\pi^2\varepsilon}\int_0^\infty\left|\int_{|x-\varepsilon|}^{|x+\varepsilon|}\frac{f(\xi)}{\xi}\,d\xi\right|^2 dx.$$

Thus (22.59) and (22.58) are equivalent to

(22.62) $$\lim_{\varepsilon\to 0}\frac{1}{\varepsilon}\int_0^\infty\left|\int_{|x-\varepsilon|}^{|x+\varepsilon|}\frac{f(\xi)}{\xi}\,d\xi\right|^2 dx = 0.$$

By the Schwarz inequality,

(22.63) $$\begin{aligned}\frac{1}{\varepsilon}\left|\int_{|x-\varepsilon|}^{|x+\varepsilon|}\frac{f(\xi)}{\xi}\,d\xi\right|^2 &\leqq \frac{1}{\varepsilon}\int_{|x-\varepsilon|}^{|x+\varepsilon|}|f(\xi)|^2\,d\xi\int_{|x-\varepsilon|}^{|x+\varepsilon|}\frac{d\xi}{\xi^2}\\ &=\left(\frac{1}{|x-\varepsilon|}-\frac{1}{|x+\varepsilon|}\right)\frac{1}{\varepsilon}\int_{|x-\varepsilon|}^{|x+\varepsilon|}|f(\xi)|^2\,d\xi.\end{aligned}$$

Thus by the boundedness of (22.55) we have uniformly in ε for all sufficiently large values of N:

(22.64) $$\begin{aligned}\frac{1}{\varepsilon}\int_{N\varepsilon}^\infty\left|\int_{|x-\varepsilon|}^{|x+\varepsilon|}\frac{f(\xi)}{\xi}\,d\xi\right|^2 dx &\leqq \int_{N\varepsilon}^\infty\frac{2\varepsilon}{x^2-\varepsilon^2}\,d\int_0^x\frac{d\eta}{\varepsilon}\int_{|\eta-\varepsilon|}^{|\eta+\varepsilon|}|f(\xi)|^2\,d\xi\\ &\leqq\int_{N\varepsilon}^\infty\frac{4\varepsilon x\,dx}{(x^2-\varepsilon^2)^2}\int_0^x\frac{d\eta}{\varepsilon}\int_{|\eta-\varepsilon|}^{|\eta+\varepsilon|}|f(\xi)|^2\,d\xi\\ &=\frac{1}{2}\int_{N\varepsilon}^\infty\frac{4\varepsilon x\,dx}{(x^2-\varepsilon^2)^2}\int_{-x-\varepsilon}^{x+\varepsilon}|f(\xi)|^2\,d\xi\int_{\xi-\varepsilon}^{\xi+\varepsilon}\frac{d\eta}{\varepsilon}\\ &\leqq \text{const.}\int_{N\varepsilon}^\infty\frac{4\varepsilon x^2\,dx}{(x^2-\varepsilon^2)^2}\\ &=\text{const.}\int_N^\infty\frac{4x^2\,dx}{(x^2-1)^2}.\end{aligned}$$

This can be made as small as we wish by taking N large enough. Moreover,

(22.65) $$\begin{aligned}\frac{1}{\varepsilon}\int_0^{N\varepsilon}\left|\int_{|x-\varepsilon|}^{|x+\varepsilon|}\frac{f(\xi)}{\xi}\,d\xi\right|^2 &= \frac{1}{\varepsilon}\int_0^{N\varepsilon}\left|\int_{|x-\varepsilon|}^{|x+\varepsilon|}\frac{1}{\xi}\,d\int_0^\xi f(\eta)\,d\eta\right|^2 dx\\ &=\frac{1}{\varepsilon}\int_0^{N\varepsilon}\left|\left[\frac{1}{\xi}\int_0^\xi f(\eta)\,d\eta\right]_{|x-\varepsilon|}^{|x+\varepsilon|}+\int_{|x-\varepsilon|}^{|x+\varepsilon|}\frac{d\xi}{\xi^2}\int_0^\xi f(\eta)\,d\eta\right|^2 dx\\ &\leqq\frac{1}{\varepsilon}\int_0^{N\varepsilon}\left|A\left(1+\log\frac{|x+\varepsilon|}{|x-\varepsilon|}\right)\right|^2 dx\\ &=A^2\int_0^N\left(1+\log\frac{|x+1|}{|x-1|}\right)^2 dx,\end{aligned}$$

where

(22.651) $$A>\left|\frac{1}{\xi}\int_0^\xi f(\eta)\,d\eta\right|\qquad(0<\xi\leqq|N+1|\varepsilon)$$

By (22.56), we may make A as small as we wish for a given N by taking ε small enough. Now let N be so large that for all ε,

$$\frac{1}{\varepsilon}\int_{N\varepsilon}^{\infty}\left|\int_{|x-\varepsilon|}^{|x+\varepsilon|}\frac{f(\xi)}{\xi}d\xi\right|^2 dx < \varepsilon_1/2, \tag{22.66}$$

and then let ε be so small that

$$\frac{1}{\varepsilon}\int_{0}^{N\varepsilon}\left|\int_{|x-\varepsilon|}^{|x+\varepsilon|}\frac{f(\xi)}{\xi}d\xi\right|^2 dx < \varepsilon_1/2. \tag{22.67}$$

It will then follow that for this and all smaller values of

$$\frac{1}{\varepsilon}\int_{0}^{\infty}\left|\int_{|x-\varepsilon|}^{|x+\varepsilon|}\frac{f(\xi)}{\xi}d\xi\right|^2 dx < \varepsilon_1. \tag{22.68}$$

Thus (22.62) follows, and Theorem XXV is established.

23. Generalized types of summability. The subject-matter of the last section leads us to interesting reflections on the notion of summability itself. The ordinary processes for summing a series are linear processes: that is, they consist in replacing the partial sums of a series by linear combinations of partial sums, and then investigating the ordinary limits of these linear combinations. This much of linearity must always remain in a definition of summability, that if two summable series are added term by term, the corresponding sums are also added, in the sense that the new sum-series will be summable to the sum of the sums corresponding to the individual series.

It is possible, however, to consider summability from another standpoint, from which linearity is not so obvious an attribute of the process. We may confine our attention to series, or rather to their sequences of partial sums, which are summable to zero. Thus a method of summability will sum to A the series whose partial sums are:

$$s_1, s_2, \cdots, s_n, \cdots \tag{23.01}$$

if it will sum to zero the series whose partial sums are:

$$s_1-A, s_2-A, \cdots, s_n-A, \cdots. \tag{23.02}$$

We may thus center our concept of summability about the notion of *null-sequence* or *null-function.* Whatever definition we choose for the class of such sequences or functions, this class should be closed additively, in the sense that the sum of any two members of the class must belong to the class. *It is not however essential that a member of the class should be characterized by the vanishing of some particular linear transform of that member.* In this sense, we are introducing a non-linear theory of summability.[62]

It is clearly a desideratum of a definition of a null-function that every function tending to a zero limit at infinity should be null. This requirement at once makes the corresponding definition of summability consistent with convergence and inclusive of it in scope. It is not satisfied by definitions of the type that assert that f is null if

$$\int_{-\infty}^{\infty} |f(x)|^p \, dx < \infty \tag{23.03}$$

but is satisfied by the definition that f is null if

$$\lim_{B\to\infty} \frac{1}{B} \int_{-B}^{B} |f(x)|^2 \, dx = 0. \tag{23.04}$$

We have already seen that if $s(u)$ is defined as in (22.23), this last statement is equivalent to the assertion that

$$\lim_{\varepsilon\to 0} \frac{1}{2\varepsilon} \int_{-\infty}^{\infty} |s(u+\varepsilon) - s(u-\varepsilon)|^2 \, du = 0. \tag{23.05}$$

This suggests an even looser definition of a null-function, according to which a function $f(x)$ is null if

$$\frac{1}{2B} \int_{-B}^{B} |f(x)|^2 \, dx \tag{23.06}$$

is bounded, and if

$$\lim_{\varepsilon\to 0} \frac{1}{2\varepsilon} \int_{-A}^{A} |s(u+\varepsilon) - s(u-\varepsilon)|^2 \, du = 0 \tag{23.07}$$

for all finite A, or even if (23.07) is true for *some* A greater than 0. The first of these definitions will make $f(x)$ null if (23.06) holds and

$$\int_{-\infty}^{\infty} K(x-\xi) f(\xi) \, d\xi \tag{23.08}$$

is null, provided that

$$\int_{-\infty}^{\infty} (1+|x|)^{3+\varepsilon} \, |K_1(x)|^2 \, dx < \infty \tag{23.09}$$

and

$$\int_{-\infty}^{\infty} K(x) \, e^{iux} \, dx \neq 0, \qquad [-\infty < u < \infty]. \tag{23.10}$$

---

[62] As far as the author knows, strong summability of various sorts is the only example of a summability process of this sort now in the literature.

The second definition will have the same property, provided only (23.10) holds for all values of u in some neighborhood of 0. All this follows from the theory developed in the last section.

These two last definitions of summability therefore fit in well with the ordinary linear definitions of summability, which define the generalized limit of a function as an expression of the form

$$\lim_{x\to\infty} \frac{\int_{-\infty}^{\infty} K(x,\xi)\, f(\xi)\, d\xi}{\int_{-\infty}^{\infty} K(x,\xi)\, d\xi} \tag{23.11}$$

on an appropriate scale of measurement. They have the great advantage of not involving any reference to a particular kernel $K(x, \xi)$. They are, however, restricted to functions for which (23.06) is bounded.

If we confine our attention to functions $f(x)$ vanishing for negative arguments, this difficulty may be overcome in its turn. We now put

$$\sigma(z) = \frac{1}{2\pi} \int_{-\infty}^{\infty} f(x)\, e^{zx}\, dx \tag{23.12}$$

throughout the half-plane $\Re(z) > u_0$: Then, by analytic continuation if necessary, $\sigma(i\nu)$ may be defined in some cases on a section of the real axis containing the origin. We put

$$s(u) = \int_0^u \sigma(iw)\, dw \tag{23.121}$$

giving σ its boundary value along the axis of imaginaries. We now define nullity as in (23.06) and (23.07). If for all z with real part between $-\epsilon$ and $u_0 + \epsilon$

$$\int_{-\infty}^{\infty} |K(x)|\, e^{zx}\, dx < \infty, \tag{23.13}$$

we have

$$\frac{1}{2\pi} \int_{-\infty}^{\infty} e^{zx}\, dx \int_{-\infty}^{\infty} K(x-\xi)\, f(\xi)\, d\xi = \sigma(z) \int_{-\infty}^{\infty} K(\xi)\, e^{z\xi}\, d\xi, \tag{23.14}$$

so that if $\int_{-\infty}^{\infty} K(\xi)\, e^{z\xi}\, d\xi$ is free from zeros and singularities in the left half plane, $f(x)$ and $\int_{-\infty}^{\infty} K(x-\xi)\, f(\xi)\, d\xi$ are null or not null simultaneously. With an appropriate definition of "analytic continuation", zeros and singularities not at the origin are of no concern, provided the function

$\int_{-\infty}^{\infty} K(\xi)\, e^{z\xi}\, d\xi$ is representible on a single-sheeted Riemann surface in the left half-plane.

It will be noted that all the definitions here given of the nullity of $f(x)$ depend on the scale chosen for x. If for all large x,

$$\frac{y'(x)}{y(A)} < \text{const. for } x < A \tag{23.15}$$

and

$$\frac{1}{2A}\int_{-A}^{A} |f(y(x))|^2\, dx \to 0 \tag{23.16}$$

then it at once follows that

$$\frac{1}{2B}\int_{-B}^{B} |f(x)|^2\, dx \to 0. \tag{23.17}$$

A similar study of the effect of a change of scale on nullity of other types would be of interest.

24. **Some unsolved problems.** In a piece of work of the ambitious length of the present, it is perhaps worth while to point out to the reader promising future directions of research. The following remarks may therefore not be amiss.

(1) The closure of the set of translations of a given function has been investigated in class L_1 and in class L_2. The methods of proof have been widely different in the two cases, but both results may be stated in a single formulation, that the set of translations of $f(x)$ is closed in the appropriate class when and only when the Fourier transform of $f(x)$, when properly defined and chosen, has no zeros. This formulation continues to constitute a reasonable proposition for L_p, where p is intermediate between 1 and 2, or even exterior to this interval. Is this proposition true? Certainly, for neither of the special cases already given is the method of proof extensible without serious modification. My own suspicion is that the general theorem is at least true for $1 \leqq p \leqq 2$.

(2) Obviously the power of Tauberian theorems in number theory has not been exhausted. Is there any Tauberian theorem which will reach from one complex ordinate to another, and enable us to handle the more refined forms of the prime number theorem?

(3) In particular, can we make a direct study of the closure of the set of all polynomials in the functions

$$e^{-\lambda(x-\xi)}\frac{d}{dx}\frac{e^{x-\xi}}{e^{e^{x-\xi}}-1}, \tag{24.01}$$

and thus attack directly the problem of the zeros of the zeta function in the critical strip?

(4) In section 22, the conditions to which $K_1(x)$ and $K_2(x)$ are subjected in Theorem XXIV are probably needlessly stringent. What is the best possible theorem in this connection?

(5) The "quasi-Tauberian" theorem of section 21 has not yet been exploited to the full. In particular, in the discussion of the relations between Riesz summability, Cesàro summability, Hölder summability, and the like, it is extremely desirable to have theorems of this type in which the kernels are not of the form $K(x-y)$, but are in some sense *nearly* of this form. Here the cruder theorems, depending solely on the use of dominant functions, are probably not difficult to elicit. On the other hand, the more refined ones will almost certainly require the full armament of Carleman's theory of singular integral equations and of the modern von Neumann-Stone[63] calculus of operators. Indeed, the time will soon come when the entire Tauberian theory must be reconsidered from the point of view of this calculus.

BIBLIOGRAPHY.

The present list of memoirs, while it cannot in the nature of things be complete, represents an attempt to bring together in one place as much as possible of the literature of Tauberian theorems. Up to 1915, we have been able to draw on the bibliography of the Cambridge tract by Hardy and M. Riesz on The General Theory of Dirichlet Series, and up to about 1924, on Smail's History and Synopsis of the Theory of Summable Infinite Processes. The bibliography also contains certain memoirs mentioned in the notes which are not of themselves Tauberian.

K. Ananda-Rau.

(1) A note on a theorem of Mr. Hardy's. Proc. Lond. Math. Soc., (2) 17 (1918), 334–336.

(2) On Lambert's series. Proc. Lond. Math. Soc., (2) 19 (1921), 1–20.

(3) On the relation between the convergence of a series and its summability by Cesàro's means. Jour. Indian Math. Soc., 15 (1923–5), 264–268.

(4) On Dirichlet's series with positive coefficients. Rendiconti di Palermo, 54 (1930), 455–462.

S. Bochner and G. H. Hardy.

(1) Note on two theorems of Norbert Wiener. Jour. Lond. Math. Soc., 1 (1926), 240–242.

E. Bortolotti.

(1) Sulle condizioni di applicabilità e di coerenza dei processi di sommazione asintotica di algoritmi infiniti. Rend. del R. Ac. di Bologna, 31 (1926–7), 33–41.

[63] Von Neumann (1); Stone (1).

S. Bosanquet.

(1) The summability of Fourier series. Math. Gaz., 15 (1931), 293-296.
(2) On the summability of Fourier series. Proc. Lond. Math. Soc., (2) 31 (1930), 144–164.

L. S. Bosanquet and E. H. Linfoot.

(1) On the zero order summability of Fourier series. Jour. Lond. Math. Soc., 6 (1931), 117–126.
(2) Generalized means and the summability of Fourier series. Quarterly Journal (Oxford series), 2 (1931), 207–229.

H. E. Bray.

(1) Elementary properties of the Stieltjes integral. Ann. of Math., 20 (1919), 176–186.

T. J. I'a. Bromwich.

(1) An introduction to the theory of infinite series. London 1908. Cf. especially p. 251
(2) On the limits of certain infinite series and integrals. Math. Ann. 65, (1908), 350–369.

Anna Caldarera.

(1) Su talune estenzioni dei criteri di convergenza di Hardy e Landau. Note e memorie di mat. (Catania), 2 (1923), 77–98.

T. Carleman.

(1) A theorem concerning Fourier series. Proc. Lond. Math. Soc., (2) 21 (1923), 483–492.

H. S. Carslaw.

(1) Introduction to the theory of Fourier series and integrals. New edition London 1921. Cf. especially pp. 150–154, 234–243.

S. Chapman.

(1) On non-integral orders of summability of series and integrals. Proc. Lond. Math. Soc., (2) 9 (1910-11), 369–409. Cf. p. 374, second note.

M. Cipolla.

(1) Criteri di convergenza riducibili a quello di Hardy-Landau. Naples Rendiconti (3a), 27 (1921), 28–37.

P. J. Daniell.

(1) A general form of integral. Ann. of Math., 19 (1917), 279–294.

P. Dienes.

(1) Sur la sommabilité de la série de Taylor. C. R., 153 (1911), 802–805.

G. Doetsch.

(1) Ein Konvergenzkriterium für Integrale. Math. Ann., 82 (1920), 68–82.
(2) Die Integrodifferentialgleichungen vom Faltungstypus. Math. Ann., 89 (1923), 192–207.
(3) Sätze von Tauberschem Charakter im Gebiet der Laplace- und Stieltjes Transformation. Berliner Berichte, Phys.-Math. Klasse (1930), 144–157.
(4) Über den Zusammenhang zwischen Abelscher und Borelscher Summabilität. Math. Ann., 104 (1931), 403–414.

P. Fatou.

(1) Séries trigonométriques et séries de Taylor. Acta Math., 30 (1906), 335-400. (Thèse, Paris 1907.)

L. Fejér.

(1) Fourierreihe und Potenzreihe. Monatshefte f. Math., 28 (1917), 64–76.

M. Fekete.

(1) Viszgálatok a Fourier sorokrol. Math. és Termész. Ért., 34 (1916).

M. Fujiwara.

(1) Über die Verallgemeinerung des Tauberschen Satzes auf Doppelreihen. Science reports of the Tôhoku national university, Sendai, Japan, 8 (1919), 43-50.

(2) Über summierbare Reihen und Integrale. Tôhoku math. jour., 15 (1919), 323-329.

(3) Ein Satz über die Borelsche Summation. Tôhoku math. jour., 17 (1920), 339-343.

J. Hadamard.

(1) Sur la distribution des zéros de la fonction $\zeta(s)$ et ses conséquences arithmétiques. Bull. Soc. Math. France, 24 (1896), 199-220.

G. H. Hardy.

(1) Researches in the theory of divergent series and divergent integrals. Quarterly journal, 35 (1904), 22-66.

(2) Theorems relating to the summability and convergence of slowly oscillating series. Proc. Lond. Math. Soc., (2) 8 (1909), 301-320.

(3) On the multiplication of Dirichlet series. Proc. Lond. Math. Soc., (2) 10 (1912), 396-405.

(4) On the summability of Fourier's series. Proc. Lond. Math. Soc., (2) 12 (1913), 365-372.

(5) An extension of a theorem on oscillating series. Proc. Lond. Math. Soc., (2), 12 (1913), 174-180.

(6) Note on Lambert's series. Proc. Lond. Math. Soc., (2) 13 (1913), 192-198.

(7) The application of Abel's method of summation to Dirichlet series. Quarterly journal, 47 (1916), 176-192.

(8) The second theorem of consistency for summable series. Proc. Lond. Math. Soc., (2) 15 (1916), 72-88.

(9) On certain criteria for the convergence of the Fourier series of a continuous function. Messenger of mathematics, 49 (1920), 149-155.

(10) The summability of a Fourier series by logarithmic means. Quarterly Journal (Oxford series), 2 (1931), 107-113.

G. H. Hardy and J. E. Littlewood.

(1) Contributions to the arithmetic theory of series. Proc. Lond. Math. Soc., (2) 11 (1913), 411-478.

(2) The relations between Borel's and Cesàro's methods of summation. Proc. Lond. Math. Soc., (2) 11 (1913), 1-16.

(3) Sur la série de Fourier d'une fonction à carré sommable. C. R. April 28, 1913.

(4) Tauberian theorems concerning power series and Dirichlet's series whose coefficients are positive. Proc. Lond. Math. Soc., (2) 13 (1913), 174-191.

(5) Some theorems concerning Dirichlet's series. Mess. math., 43 (1914), 134-147.

(6) Theorems concerning the summability of series by Borel's exponential method. Rendiconti di Palermo, 41 (1916), 36-53.

(7) Sur la convergence des séries de Fourier et des séries de Taylor. C. R. Dec. 24, 1917.

(8) The Riemann zeta-function and the theory of the distribution of primes. Acta Math., 41 (1918), 119-196.

(9) On the Fourier series of a bounded function. Proc. Lond. Math. Soc., (2) 17 (1918), xiii-xv. (Abstract).

(10) Abel's theorem and its converse. Proc. Lond. Math. Soc., (2) 18 (1920), 205-235.

(11) On a Tauberian theorem for Lambert's series, and some fundamental theorems in the analytic theory of numbers. Proc. Lond. Math. Soc., (2) 19 (1921), 21–29.
(12) Solution of the Cesàro summability problem for power-series and Fourier series. Math. Ztschr., 19 (1923), 67–96.
(13) Abel's theorem and its converse (II). Proc. Lond. Math. Soc., (2) 22 (1924), 254–269.
(14) The allied series of a Fourier series. Proc. Lond. Math. Soc., 24 (1925), 211–246.
(15) The strong summability of Fourier series. Proc. Lond. Math. Soc., (2) 26 (1926), 273–286.
(16) Elementary theorems concerning power series with positive coefficients and moment-constants of positive functions. Crelle, 157 (1926), 141–158.
(17) A further note on the converse of Abel's theorem. Proc. Lond. Math. Soc., (2) 25 (1926), 219–236.
(18) Notes on the theory of series (VII): on Young's convergence criterion for Fourier series. Proc. Lond. Math. Soc., (2) 28 (1928) 301–311.
(19) Notes on the theory of series (XI): on Tauberian theorems. Proc. Lond. Math. Soc., (2) 30 (1929), 23–37.
(20) Notes on the theory of series (III): on the summability of the Fourier series of a nearly continuous function. Proc. Camb. Phil. Soc., 23 (1927), 681–684.
(21) The equivalence of certain integral means. Proc. Lond. Math. Soc. records, (2) 22 (1924), xl–xliii.

G. H. Hardy and M. Riesz.

(1) The general theory of Dirichlet series. Cambridge Tract in Mathematics and Mathematical Physics, No. 18. Cambridge 1915.

G. H. Hardy and E. C. Titchmarsh.

(1) Self-reciprocal functions. Quarterly journal (Oxford series), 1 (1930), 196–232.

W. A. Hurwitz.

(1) A trivial Tauberian theorem. Bull. Am. Math. Soc., 32 (1926), 77–82.

S. Ikehara.

(1) An extension of Landau's theorem in the analytic theory of numbers. Journal of Math. and Phys. of the Mass. Inst. of Technology, 10 (1931), 1–12.

S. Izumi.

(1) A generalization of Tauber's theorem. Proc. Imperial Acad. Japan, 5 (1929), 57–59.

M. Jacob.

(1) Über den Eindeutigkeitssatz in der Theorie der verallgemeinerten trigonometrischen Integrale. Math. Ann., 100 (1926–7), 278–294.
(2) Über ein Theorem von Bochner-Hardy-Wiener. Jour. Lond. Math. Soc., 3 (1928), 182–187.

J. Karamata.

(1) Sur le mode de croissance régulière des fonctions. Mathematica (Cluj, Roumania), 4 (1930), 38–53.
(2) Über die Hardy-Littlewoodschen Umkehrungen des Abelschen Stetigkeitssatzes. Math. Zeitschr., 32 (1930), 319–320.
(3) Neuer Beweis und Verallgemeinerung einiger Tauberian-Sätze. Math. Zeitschr., 33 (1931), 294–300.
(4) Neuer Beweis und Verallgemeinerung der Tauberschen Sätze, welche die Laplacesche und Stieltjessche Transformation betreffen. Crelle, 164 (1931), 27–40.

A. Kienast.

(1) Extension to other series of Abel's and Tauber's theorems on power series. Proc. Lond. Math. Soc., (2) 25 (1926), 45–52.

K. Knopp.

(1) Neuere Untersuchungen in der Theorie der divergenten Reihen. Jahresber. d. deutschen Mathematikervereinigung, 32 (1923), 43–67.

(2) Theorie und Anwendung der unendlichen Reihen. Berlin, Springer, 1922. Chap. XIII.

E. Landau.

(1) Über die Konvergenz einiger Klassen von unendlichen Reihen am Rande des Konvergenzgebietes. Monatsh. f. Math. und Phys., 18 (1907), 8–28.

(2) Handbuch der Verteilung der Primzahlen. Leipzig 1909, 2v.

(3) Über die Bedeutung einiger neuerer Grenzwertsätze der Herren Hardy und Axer. Prace mat.-Fiz., 21 (1910), 97–177.

(4) Über einen Satz des Herrn Littlewood. Rendiconti di Palermo, 35 (1913), 265–276.

(5) Ein neues Konvergenzkriterium für Integrale. Münch. Sitzber., 43 (1913), 461–467.

(6) Darstellung und Begründung einiger neuerer Ergebnisse der Funktionentheorie. Berlin 1916, 2nd ed., 1930.

(7) Sobre los números primos en progressión aritmética. Rev. Mat. Hispano-Americana, 4 (1923), 1–16, 33–44.

(8) Vorlesungen über Zahlentheorie. Leipzig 1927, 3v.

H. Lebesgue.

(1) Recherches sur la convergence des séries de Fourier. Math. Ann., 61 (1905), 251–280.

P. Lévy.

(1) Sur les conditions d'application et sur la régularité des procédés de sommation des séries divergentes. Bull. Soc. Math. de France, 54 (1926), 1–25.

S. B. Littauer.

(1) On a theorem of Jacob. Jour. Lond. Math. Soc., 4 (1929), 226–231.

(2) A new Tauberian theorem with application to the summability of Fourier series and integrals. Jour. Math. Phys. Mass. Insti. Tech., 8 (1928), 216–234.

J. E. Littlewood.

(1) On the converse of Abel's theorem on power series. Proc. Lond. Math. Soc., (2) 9 (1910), 434–448.

H. J. Mellin.

(1) Die Theorie der asymptotischen Reihen vom Standpunkte der reziproken Funktionen und Integrale. Ann. Ac. Sci. Fennicae, (A) 18 (1922), No. 4, 108 p.

G. Mignosi.

(1) Inversione d'un teorema sul rapporto delle medie (Cp) di due serie. Naples Rendiconti (3a), 27 (1921), 17–28.

L. Neder.

(1) Über Taubersche Bedingungen. Proc. Lond. Math. Soc., (2) 23 (1925), 172–184.

J. von Neumann.

(1) Allgemeine Eigenwerttheorie Hermitescher Funktionaloperatoren. Math. Ann., 102 (1930), 49–131.

R. E. A. C. Paley.

(1) On the Cesàro summability of Fourier series and allied series. Proc. Camb. Phil. Soc., 26 (1930), 173–203.

M. Plancherel.

(1) Contribution à l'étude de la répresentation d'une fonction arbitraire par des intégrales définies. Rendiconti di Palermo, 30 (1910), 289–335.

A. Pringsheim.

(1) Über das Verhalten von Potenzreihen auf dem Konvergenzkreise. Münch. Ber., 30 (1900), 43–100.

(2) Über die Divergenz gewisser Potenzreihen an der Konvergenzgrenze. M. Ber., 31 (1901), 505–524.

(3) Über eine Konvergenzbedingung für unendliche Reihen, die durch iterierte Mittelbildung reduzibel sind. M. Ber., 50 (1920), 275–284.

M. Riesz.

(1) Une méthode de sommation équivalente à la méthode des moyennes arithmétiques. C. R., 12 June 1911.

(2) Über einen Satz des Herrn Fatou. Crelle, 140 (1911) 89–99.

R. Schmidt.

(1) Über das Borelsche Summierungsverfahren. Schriften der Königsberger gelehrten Gesellschaft, 1 (1925), 202–256.

(2) Über divergente Folgen und lineare Mittelbildungen. M. Ztsch., 22 (1925), 89–152.

W. Schnee.

(1) Über Dirichletsche Reihen. Rend. di Palermo, 27 (1909), 87–116.

J. Schur.

(1) Über lineare Transformationen in der Theorie der unendlichen Reihen. Crelle, 151 (1921), 79–111.

L. L. Smail.

(1) History and synopsis of the theory of summable infinite processes. U. of Oregon publ., 2 (Feb. 1925), No. 8.

M. H. Stone.

(1) Linear transformations in Hilbert space III. Operational methods and group theory. Proc. Mat. Acad. Sci., 16 (1930), 172–175.

O. G. Sutton.

(1) On a theorem of Carleman. Proc. Lond. Math. Soc., (2) 23 (1925), xlvii–li.

O. Szász.

(1) Verallgemeinerung und neuer Beweis einiger Sätze Tauberischer Art. Münch. Ber., (1929), 325–340.

(2) Über einen Satz von Hardy und Littlewood. Berliner Berichte 1930, 4p.

(3) Über die Approximation stetiger Funktionen durch gegebene Funktionenfolgen. Math. Ann., 104 (1930), 155–160.

(4) Abel hatványsortételével kapcsolatos újabb vizsgálatokról. Mat. és Phys. Lapok., 36 (1929), 10–22.

(5) Über Sätze Tauberscher Art. Jahresb. d. Deutschen Mathematikervereinigung, 39 (1930), 28–31.

(6) Über einige Sätze von Hardy und Littlewood. Göttinger Nachrichten (1930), Fachgruppe I, 315–333.

(7) Verallgemeinerung eines Littlewoodschen Satzes über Potenzreihen. J. L. M. S., 3 (1930), 254–262.

(8) Über Dirichletsche Reihen an der Konvergenzgrenze. Rend. del Cong. Math. di Bologna, (1928), 269–276.

A. Tauber.

(1) Ein Satz aus der Theorie der unendlichen Reihen. Monatshefte f. Math., 8 (1897). 273–277.

E. C. Titchmarsh.

(1) A theorem on infinite products. Jour. Lond. Math. Soc., 1 (1926), 35–37.
(2) On integral functions with real negative zeros. Proc. Lond. Math. Soc., (2) 26 (1927), 185–200.
(3) A contribution to the theory of Fourier transforms. Proc. Lond. Math. Soc., (2) 23 (1924), 279–289.

C. de la Vallée Poussin.

(1) Récherches analytiques sur la théorie des nombres premiers. Première partie. Ann. Soc. Sci. Brux., 20: 2 (1896), 183–256.

S. Verblunsky.

(1) The convergence of singular integrals. Proc. Camb. Phil. Soc., 26 (1930), 312–322.
(2) On summable trigonometric series. Proc. Lond. Math. Soc., (2) 31 (1930), 370–386.
(3) On the limit of a function at a point. Proc. Lond. Math. Soc., (2) 32 (1931), 163–200.
(4) A uniqueness theorem for trigonometric series. Quarterly Journal (Oxford series), (2) (1931), 81–90.

T. Vijayaraghavan.

(1) A theorem concerning the summability of a series by Borel's method. Proc. Lond. Math. Soc., (2) 27 (1928), 316–326.
(2) A Tauberian theorem. Jour. Lond. Math. Soc., 1 (1926), 113–230.

N. Wiener.

(1) The harmonic analysis of irregular motion. Jour. Math. and Phys. Mass. Inst. Tech., 5 (1925), 99–122.
(2) Une méthode nouvelle pour la démonstration des théorèmes de M. Tauber. C. R., 184 (1927), 793–795.
(3) On a theorem of Bochner and Hardy. Jour. Lond. Math. Soc., 2 (1927), 118–123.
(4) A new method in Tauberian theorems. Jour. Math. Phys. M. I. T., 7 (1928), 161–184.
(5) A type of Tauberian theorem applying to Fourier series. Proc. Lond. Math. Soc., (2) 30 (1929), 1–8.
(6) The spectrum of an arbitrary function. Proc. Lond. Math. Soc., (2) 27 (1928), 487–496.
(7) Generalized harmonic analysis. Acta. Math., 55 (1930), 117–258.
(8) The operational calculus. Math. Ann., 95 (1926), 557–584.

A. Wintner.

(1) Untersuchungen über Funktionen großer Zahlen. (Erste Mitteilung) M. Ztschr., 28 (1928), 416–429.

W. H. Young.

(1) On the determination of the summability of a function by means of its Fourier constants. Proc. Lond. Math. Soc., (2) 12 (1912), 71–88.

THE M.I.T. PAPERBACK SERIES

1 **Computers and the World of the Future** edited by Martin Greenberger

2 **Experiencing Architecture** by Steen Eiler Rasmussen

3 **The Universe** by Otto Struve

4 **Word and Object** by Willard Van Orman Quine

5 **Language, Thought, and Reality** by Benjamin Lee Whorf

6 **The Learner's Russian-English Dictionary** by B. A. Lapidus and S. V. Shevtsova

7 **The Learner's English-Russian Dictionary** by S. Folomkina and H. Weiser

8 **Megalopolis** by Jean Gottmann

9 **Time Series** by Norbert Wiener

10 **Lectures on Ordinary Differential Equations** by Witold Hurewicz

11 **The Image of the City** by Kevin Lynch

12 **The Sino-Soviet Rift** by William E. Griffith

13 **Beyond the Melting Pot** by Nathan Glazer and Daniel Patrick Moynihan

14 **A History of Western Technology** by Friedrich Klemm

15 **The Dawn of Astronomy** by Norman Lockyer

16 **Information Theory** by Gordon Raisbeck

17 **The Tao of Science** by R. G. H. Siu

18 **A History of Civil Engineering** by Hans Straub

19 **Ex-Prodigy** by Norbert Wiener

20 **I Am a Mathematician** by Norbert Wiener

21 **The New Architecture and the Bauhaus** by Walter Gropius

22 **A History of Mechanical Engineering** by Aubrey F. Burstall

23 **Garden Cities of To-Morrow** by Ebenezer Howard

24 **Brett's History of Psychology** edited by R. S. Peters

25 **Cybernetics** by Norbert Wiener

26 **Biological Order** by André Lwoff

27 **Nine Soviet Portraits** by Raymond A. Bauer

28 **Reflexes of the Brain** by I. Sechenov

29 **Thought and Language** by L. S. Vygotsky

30 **Chinese Communist Society: The Family and the Village** by C. K. Yang

31 **The City: Its Growth, Its Decay, Its Future** by Eliel Saarinen

32 **Scientists as Writers** edited by James Harrison

33 **Candidates, Issues, and Strategies: A Computer Simulation of the 1960 and 1964 Presidential Elections** by I. de S. Pool, R. P. Abelson, and S. L. Popkin

34 **Nationalism and Social Communication** by Karl W. Deutsch

35 **What Science Knows About Life: An Exploration of Life Sources** by Heinz Woltereck

36 **Enzymes** by J. B. S. Haldane

37 **Universals of Language** edited by Joseph H. Greenberg

38 **The Psycho-Biology of Language: An Introduction to Dynamic Philology** by George Kingsley Zipf

39 **The Nature of Metals** by Bruce A. Rogers

40 **Mechanics, Molecular Physics, Heat, and Sound** by R. A. Millikan, D. Roller, and E. C. Watson

41 **North American Trees** by Richard J. Preston, Jr.

42 **God and Golem, Inc.** by Norbert Wiener

43 **The Architecture of H. H. Richardson and His Times** by Henry-Russell Hitchcock

44 **Toward New Towns for America** by Clarence Stein

45 **Man's Struggle for Shelter in an Urbanizing World** by Charles Abrams

46 **Science and Economic Development** by Richard L. Meier

47 **Human Learning** by Edward Thorndike

48 **Pirotechnia** by Vannoccio Biringuccio

49 **A Theory of Natural Philosophy** by Roger Joseph Boscovich

50 **Bacterial Metabolism** by Marjory Stephenson

51 **Generalized Harmonic Analysis** and **Tauberian Theorems** by Norbert Wiener

52 **Nonlinear Problems in Random Theory** by Norbert Wiener

53 **The Historian and the City** edited by Oscar Handlin and John Burchard

54 **Planning for a Nation of Cities** edited by Sam Bass Warner, Jr.

55 **Silence** by John Cage

56 **New Directions in the Study of Language** edited by Eric H. Lenneberg

57 **Prelude to Chemistry** by John Read

58 **The Origins of Invention** by Otis T. Mason

59 **Style in Language** edited by Thomas A. Sebeok

60 **World Revolutionary Elites** edited by Harold D. Lasswell and Daniel Lerner